570

Syn in

ADVANCED
BIOLOGY

Hodder & Stoughton

A MEMBER OF THE HODDER HEADLINE GROUP

Acknowledgements

I would like to thank Edexcel, OCR and AQA for allowing me to reproduce some synoptic questions. I should make it clear that the exam boards accept no responsibility whatsoever for the accuracy of the method of working in the answers given.

Dedication
To Helen and Isabelle.
I love you both very much.

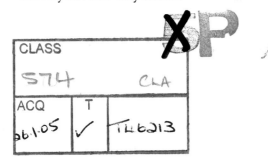
Orders: please contact Bookpoint Ltd, 130 Milton Park, Abingdon, Oxon OX14 4SB. Telephone: (44) 01235 827720. Fax: (44) 01235 400454. Lines are open from 9.00–6.00, Monday to Saturday, with a 24 hour message answering service. You can also order through our website www.hodderheadline.co.uk

British Library Cataloguing in Publication Data
A catalogue record for this title is available from the British Library

ISBN 0 340 80322 3

First Published 2001
Impression number   10 9 8 7 6 5
Year   2007 2006 2005 2004

Copyright © 2001 Alan Clamp

Typeset by Tech Set Ltd
Printed in Great Britain for Hodder & Stoughton Educational, a division of Hodder Headline Ltd., 338 Euston Road, London NW1 3BH by J. W. Arrowsmith Ltd., Bristol.

CONTENTS

CHAPTER ONE

INTRODUCTION

Why do you need this book?

In September 2001, students in their second year of a GCE Advanced Biology (or Human Biology) course will begin to study the new A2 qualification for the first time. A2 Biology represents a significant 'step up' from the AS qualification for a number of reasons:

★ The AS qualification is designed to be at a standard intermediate between GCSE and A-level. In other words, you should be able to reach AS standard in one year after GCSE. A2 Biology, however, is set at a higher standard, such that AS and A2 combined represents an A-level qualification.

★ AS units generally contain simpler, more accessible topics than A2 units. For example, at AS you may learn information about molecules and enzymes, but at A2 you would be expected to use this information to explain complex metabolic pathways, such as aerobic respiration.

★ Question styles in the unit tests will be different at AS and A2. AS questions tend to be more structured and more easily understood than those at A2. For example, an AS question on ecology may ask you to fill-in the gaps (using the appropriate word or words) in a short passage about the nitrogen cycle; an A2 question on this topic may require you to analyse some fairly complicated experimental data and to explain the results of a study, or to write an essay on nutrient cycles.

★ The assessment objectives of the specification (syllabus) are weighted differently at AS and A2. Assessment objective 1 (AO1) covers the fairly simple skills of knowledge and understanding, and represents about 60% of the marks on the written papers for AS, but only about 35% of the marks on the written papers for A2. The more complicated skills of *applying* knowledge and understanding to new situations, plus analysis, synthesis and evaluation of information, are covered by assessment objective 2 (AO2). This represents about 40% of the marks on the written papers for AS and 25% of the marks on the written papers for A2. The remainder (40%) of the marks on the A2 papers are for assessment objective 4 (AO4), which covers *synoptic skills*.

★ There is no synoptic assessment, therefore, in AS Biology, but almost half the marks on the written papers for A2 Biology come from (potentially very difficult) synoptic questions.

In summary, the new A2 unit tests for all awarding bodies have 40% of the marks for synoptic skills (taking both written and practical/coursework papers into account). Even when considered as 20% of the whole A-level, this still represents more marks than any single 'content-based' unit. Synoptic papers were not done well by the vast majority of candidates taking the 'old' A-level examinations and the much greater emphasis on synoptic skills in Curriculum 2000 will mean that you will have to develop the skills required if you are to succeed. The aim of this book, therefore, is to help you to develop the synoptic skills required in A-level Biology and to provide invaluable advice and guidance about synoptic assessment in the written examinations.

How to use this book

Having decided that you *do* need this book, it may be useful to offer some advice as to how to use it most effectively.

1 Review the advice on study skills and revision strategies given in the rest of this introduction. If you can develop good study skills, you will find that learning A-level Biology becomes much easier, including the development of synoptic skills.

2 Make sure that you are clear about the nature of synoptic skills and the manner in which your awarding body (exam board, e.g. AQA(A), AQA(B), Edexcel or OCR) incorporates synopsis into the A-level Biology specification (see pages 7–8).

3 Gradually develop and improve your synoptic skills using the various strategies suggested in Chapter 3, Making Connections: Developing Synoptic Skills (see pages 9–32).

4 Make sure that you know how these synoptic skills will be assessed by your awarding body, i.e. *where* they are assessed (which examination paper or coursework task) and *how* they are assessed (see pages 33–34).

5 Analyse a variety of synoptic question styles to see what makes a question synoptic and how best to tackle these difficult questions (see pages 35–59).

6 Practise answering synoptic examination questions under timed conditions in the mock examination papers given towards the back of this book (see pages 60–75). Compare your answers to those provided in the mark schemes, noting the useful advice from the examiner.

Study skills and revision strategies

Before worrying about synoptic skills, you need to develop good study skills if they are to be successful. This section of the introduction provides advice and guidance on how to study A-level Biology and suggests some strategies for effective revision.

Organising your notes

Biology students usually accumulate a large quantity of notes and it is useful to keep this information in an organised manner. The presentation of notes is important; good notes should always be clear and concise. You could try organising your notes under headings and subheadings, with key points highlighted using capitals, italics or colour. Numbered lists are useful, as are tables and diagrams. It is a good idea to file your notes in specification order, using a consistent series of informative headings, as illustrated below.

> # UNIT 4 (Respiration and coordination)
> ## Regulation of the internal environment
>
> *Nervous coordination in mammals*
>
> There are several important differences between nervous and hormonal coordination…

After the lessons, it is a good idea to check your notes using your textbook(s) and fill in any gaps in the information. Make sure you go back and ask the teacher if you are unsure about anything, especially if you find conflicting information in your class notes and textbook.

Organising your time

When trying to organise your time, it is a good idea to make a revision timetable. This should allow enough time to cover all the material, but also be realistic. For example, it is useful to leave some time at the end of the timetable, just before the unit test, to catch up on time lost, for example through illness. You may not be able to work for very long at a single session – probably no more than one hour without a short break of 10–15 minutes. It is also useful to use spare moments, such as when waiting for a bus or train, to do short snippets of revision. These 'odd minutes' can add up to many hours.

Improving your memory

There are several things you can do to improve the effectiveness of your memory for biological information. Organising the material will help, especially if you use topic headings, numbered lists and diagrams. Repeatedly reviewing your notes will also be useful, as will discussing topics with teachers and other students. Finally using mnemonics (memory aids), such as **A**rteries carry blood **A**way from the heart, can make a big difference.

Revision strategies

To revise a topic effectively you should work carefully through your notes, using a copy of the specification to make sure you have not missed anything out. Summarise your notes to the bare essentials, using the tips given on note-making above. Finally, use the content guidance and mock examinations in this book, discussing any difficulties with your teachers or fellow students.

In many ways, a student should prepare for a unit test like an athlete prepares for a major event, such as the Olympic Games. The athlete will train every day for weeks or months before the event, practising the required skills in order to achieve the best performance on the day. So it is with test preparation: everything you do should contribute to your chances of success in the unit test. The following points summarise some of the strategies that you may wish to use to make sure that your revision is as effective as possible.

★ Use a revision timetable.

★ Ideally, revise in a quiet room, sitting at a desk or table, with no distractions.

★ Test yourself regularly to assess the effectiveness of your revision.

★ Practise previous test questions to highlight gaps in your knowledge and understanding and to improve your technique.

★ Active revision is much better than simply reading over material. Discuss topics, summarise notes and use the mock tests included in this book to increase the effectiveness of your revision.

The unit tests

There are a number of terms commonly used in unit tests. It is important that you understand the meaning of each of these terms and that you answer the questions appropriately.

★ CALCULATE – carry out a calculation, showing your working and providing the appropriate units.

★ COMPARE – point out similarities *and* differences.

★ DEFINE – give a statement outlining what is meant by a particular term.

★ DESCRIBE – provide an accurate account of the main points, an explanation is *not* necessary.

★ DISCUSS – describe and evaluate, putting forward the various opinions on a topic.

★ DISTINGUISH BETWEEN – point out differences only.

★ EXPLAIN – give reasons, with reference to biological facts. A description is *not* required.

★ OUTLINE – give a brief account.

★ SIGNIFICANCE – the relevance of an idea or observation.

★ STATE – give a concise, factual answer (also applies to GIVE or NAME).

★ SUGGEST – use biological knowledge to put forward an appropriate answer in an unfamiliar situation.

★ WHAT/WHY/WHERE – direct questions requiring concise answers.

Whatever the question style, you must read the question *very carefully*, underline key words or phrases, think about your response and allocate time according to the number of marks available. Further advice and guidance on answering test questions is provided in Chapter 6, Mock Examinations, at the end of this book.

Structured questions

These are short-answer questions that may require a single-word answer, a short sentence, or a response amounting to several sentences. Answers should be clear, concise and to the point. The marks allocated and the space provided for the answer usually give an indication of the amount of detail required. Typical question styles include:

★ naming parts on diagrams

★ filling in gaps in a prose passage

★ completing tables and tick-boxes

★ plotting graphs

★ performing calculations

★ interpreting experimental data.

Free-prose questions

These questions enable you to demonstrate the depth and breadth of your biological knowledge, as well as your ability to communicate scientific ideas in a concise and clear manner. The following points should help you to perform well when answering free-prose questions.

★ Make your points clearly and concisely, illustrating with examples where appropriate.

★ Try to avoid repetition and keep the answer relevant (refer back to the question).

★ The points you make should cover the *full range* of the topics addressed in the question.

★ Use diagrams only if appropriate and where they make a useful contribution to the quality of your answer.

★ Spend the appropriate amount of time on the question (proportional to the marks available).

Essays

Essay questions test your ability to describe and explain biological systems and processes and also your understanding of biological principles and concepts. In the essay, you may be required to sustain an argument, to present evidence for and against a statement and to show that you are aware of the implications and applications of modern biology.

To score high marks in an essay, you need to consider the points made about free-prose questions above and also make sure that you develop a coherent argument and demonstrate high quality written communication.

The day of the unit test

On the day of the test, make sure that you have:

★ two or more blue/black pens, and two or more pencils

★ your calculator plus spare batteries

★ a watch to check the time

★ a ruler and an eraser.

Read each question very carefully so that your answers are appropriate. Make sure that you write legibly (you cannot be given marks if the examiner cannot read what you have written) and try to spell scientific terms accurately. If you need more room for your answer, look for space at the bottom of the page, the end of the question or after the last question, or use supplementary sheets. If you use these spaces, or sheets, alert the examiner by adding 'continued below', or 'continued on page X'.

Time is often a problem. Make sure that you know how long you have got for the whole test and how many questions you have to do in this time. You could use the number of minutes per mark to work out approximately how long you have for each question.

Do not write out the question, but try to make a number of valid points that correspond to the number of marks available. If you get stuck, make a note of the question number and move on. Later, if you have time, go back and try that difficult question again. Finally, it is a good idea to leave a few minutes at the end to check through the paper, correcting any mistakes or filling in any gaps.
Good luck!

CHAPTER TWO

WHAT ARE SYNOPTIC SKILLS?

The A-level Biology specifications of the three major awarding bodies (AQA, Edexcel and OCR) state that students should be able to:

★ bring together principles and concepts from different areas of biology and apply them in a particular context

★ use biological skills in contexts which bring together different areas of the subject

★ express ideas clearly and logically, using appropriate specialist vocabulary.

It will be useful to consider each of these skills in turn.

Students should be able to bring together principles and concepts from different areas of biology and apply them in a particular context.

This statement means that you should be able to use your knowledge of facts (e.g. the carbohydrate found in DNA is deoxyribose) and general principles (e.g. osmosis is the movement of water molecules from a region of higher water potential [Ψ] to a region of lower water potential, across a partially permeable membrane) to answer a question set in a particular context. Non-synoptic questions may focus exclusively on one topic (which is contained within a single unit), such as the structure of DNA. Synoptic questions, however, require you to make connections between the facts or principles in different units. For example, an essay on the roles of carbohydrates in living organisms may require you to mention their structural roles (e.g. deoxyribose in DNA), as well as their transport roles (e.g. sucrose in the phloem of flowering plants), with this information coming from different units.

Most A-level Biology students are good at learning 'facts' and therefore just need to develop their synoptic skills by making connections between these facts (see Chapter 3, Making Connections, on pages 9–32). Many of these students, however, are less clear about the key 'principles', 'concepts' or 'themes' in biology and a few examples of these are listed below.

★ Different levels of organisation (atom–molecule–macromolecule–cell–tissue–organ–organ system–organism).

★ The relationship between structure and function.

★ The importance of surface area to volume ratio.

★ The roles of biological molecules.

★ Exchanges of materials with the environment.

★ Transport across membranes and within multicellular organisms.

★ Natural selection and adaptations to the environment.

★ Energy flow through ecosystems.

★ Homeostasis and the principles of negative feedback.

★ Nervous and hormonal coordination.

Students should be able to use biological skills in contexts which bring together different areas of the subject.

Again, you need to be able to bring together different areas of the subject (see Chapter 3, Making Connections, on pages 9–32), but the emphasis in this statement is on *skills*. The skills you need are as follows:

★ know and understand biological information

★ understand the ethical, social, economic, environmental and technological implications and applications of biology

★ describe, explain and interpret information in terms of biological principles and concepts

★ interpret data presented in various forms

★ apply biological principles and concepts in solving problems in unfamiliar situations

★ plan experiments and discuss the results of experiments.

You also need to demonstrate certain mathematical skills, such as: calculating percentages and rates of change; finding the mean of a set of data; applying and interpreting statistical tests; plotting graphs; and using algebraic formulae.

Students should be able to express ideas clearly and logically, using appropriate specialist vocabulary.

All A-level examinations assess your ability to communicate. Although there are no separate marks for this skill (as there are at GCSE), poor quality English will make it harder for you to express your biological knowledge in a clear fashion and therefore you will still lose marks. Handwriting must be legible (you cannot earn marks if the examiner cannot read your answers) and spelling must be accurate, especially for terms which look similar but have very different meanings, such as 'glycogen' and 'glucagon'. Grammar is important in all questions requiring an answer written in sentences, where poor punctuation may lose you marks by making the answer ambiguous. You should also use specialist vocabulary whenever possible, e.g. using the term 'trachea' or 'bronchus' instead of 'windpipe'. Finally, some essay questions *do* award specific marks for the quality of communication. In order to achieve maximum marks in such as essay, you should ensure that you:

★ have an introductory paragraph outlining the nature of the topic under discussion

★ write a balanced essay, covering most of the key ideas in a number of paragraphs organised in a logical manner

★ finish the essay with a concluding paragraph which provides an overview of what you have written, together with any conclusion(s).

CHAPTER THREE

MAKING CONNECTIONS: DEVELOPING SYNOPTIC SKILLS

Brainstorming

Many A-level Biology courses are (understandably) taught unit-by-unit and examined in the same way. It can, therefore, sometimes be difficult to see the connections between topics in different units. Brainstorming is the art of seeking solutions to problems or developing new ideas by spontaneous suggestions. When you brainstorm on a particular topic, such as 'the functions of proteins', you should produce a long list of topics or facts associated with protein function. These can be grouped into themes and the various topics/facts can be linked by making a 'spider diagram' (see Figure 3.1). It is very unlikely that the topics or facts that you have thought of will be confined to a single unit within the A-level Biology specification and so you are already thinking synoptically and (most importantly) 'making connections'.

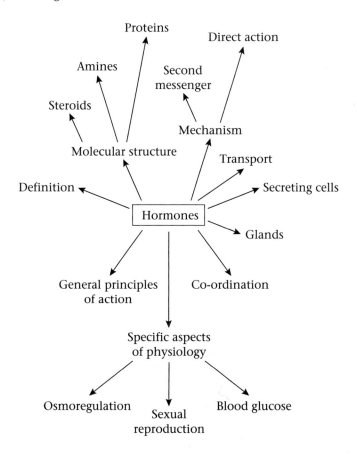

Figure 3.1

To brainstorm, you should pick a topic from within the specification, such as 'hormones'. Then brainstorm other topics that are connected to this central idea (it may help to look at the specification), trying to produce a coherent spider diagram. This has been done for 'hormones' in Figure 3.1.

Note that Figure 3.1 includes a view of hormones from a molecular, cellular and physiological viewpoint. Of course, depending on the specification you are studying, this may not be synoptic, because all the relevant information is contained within one unit (although this is not likely). You may, therefore, wish to go one step further and identify the relevant units in the diagram itself. This has been done for hormones in Figure 3.2, according to the Edexcel specification.

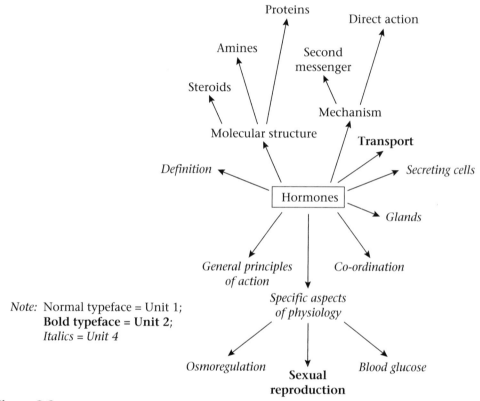

Figure 3.2

Instead of using a specific topic from the specification, you may want to use a theme, such as 'transport', or even a general principle, such as 'structure is related to function' as the focus for your brainstorm. This has been done for transport and the importance of molecular shape in Figures 3.3 and 3.4 respectively.

This process of making connections between units by looking for overlapping topics or common themes/concepts is extremely useful for developing your synoptic skills and it is worth practising. You might want to work with a friend (or group of friends) to produce and swap these biological spider diagrams, but remember that actually making the connections yourself is invaluable.

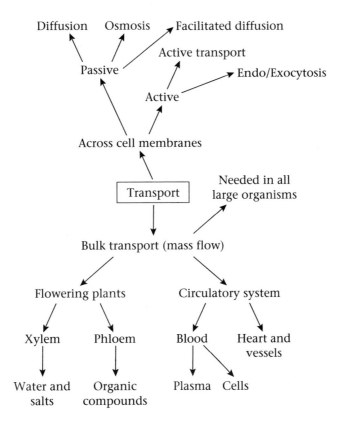

Figure 3.3

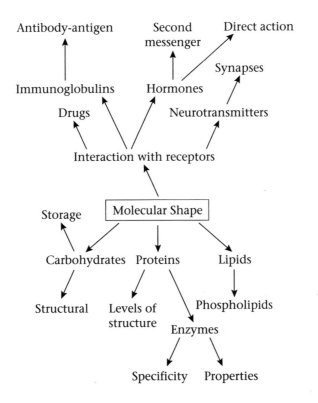

Figure 3.4

Try producing brainstorms/spider diagrams for

a) *enzymes* and

b) *gas exchange in plants and animals.*

Suggested answers (although there are no 'right' answers to this exercise) are provided in Chapter 7, Making Connections: answers to exercises, on pages 76–83.

Analysing data

Analysing data is synoptic because it often requires you to bring together information from different areas of the specification in order to explain experimental results. Furthermore, it also represents an opportunity for you to use the biological skills of plotting graphs, taking readings, identifying trends, selecting and using information from tables or graphs, and performing calculations. As a result, the analysis of data satisfies the first two criteria for synopsis given on page 7.

Constructing tables

The results of any biological experiment should be recorded as the investigation is taking place. This leaves you with a list of raw results that need to be presented in a suitable format that can be understood and used to draw conclusions about the experiment. A standard way of summarising data is to construct a table.

The headings for each of the columns in a table should indicate what observations or measurements have been made and the nature of the units used. It should be possible to determine exactly how the experiment has been carried out from the information given in the headings of a table of results. Table 3.1 is an example of a results table for an experiment on the breakdown of starch by amylase.

Temperature / °C	Time (t) taken for starch to disappear / min				Rate of reaction (1/t × 100)
	First reading	Second reading	Third reading	Mean	
5	42	47	53	47.3	2.1
15	21.5	26.5	27.5	25.2	4.0
25	13	15.5	15	14.5	6.9
35	6	8.5	8	7.5	13.3
45	3.5	5	6	4.8	20.8
55	12	15.5	17	14.8	6.8

Table 3.1

EXERCISE 2

Construct a table to summarise the following data.

An investigation was carried out to find the effect of altitude on the number of red blood cells in men and women. The results (expressed as number of red blood cells $\times 10^{12}$ per dm^3) are summarised below.

LOW ALTITUDE (*men*): 5.5, 5.4, 5.1, 5.6, 4.9, 5.0, 5.5, 5.2, 5.1 and 5.6.

LOW ALTITUDE (*women*): 4.9, 4.8, 4.7, 4.6, 4.8, 4.7, 4.9, 4.5, 4.4 and 4.6.

HIGH ALTITUDE (*men*): 5.5, 5.8, 5.7, 5.9, 5.3, 5.8, 5.7, 5.5, 5.6 and 5.7.

HIGH ALTITUDE (*women*): 5.0, 5.1, 4.9, 4.7, 4.9, 4.9, 5.1, 5.2, 5.0 and 4.9.

A suggested answer is provided in Chapter 7, Making Connections: answers to exercises, on pages 76–83.

The requirement to construct a table is rare in written examinations, but may form part of the synoptic marks for those awarding bodies that use coursework as part of synoptic assessment (AQA(A), AQA(B) and OCR).

Plotting graphs

Most experiments generate quantitative data (numerical results) which can be plotted as a graph, histogram or bar chart, so that the trends or patterns in the data show up clearly. The nature of the data collected dictates the choice of graph type.

★ **Line graph:** used when the relationship between two variables can be represented as a continuum, e.g. the effect of temperature on the activity of amylase (Figure 3.5).

★ **Bar chart:** used to show the frequency distribution of a discrete variable (one which can be divided into non-overlapping categories), e.g. the percentage of a population having blood group A, B, AB or O (Figure 3.6).

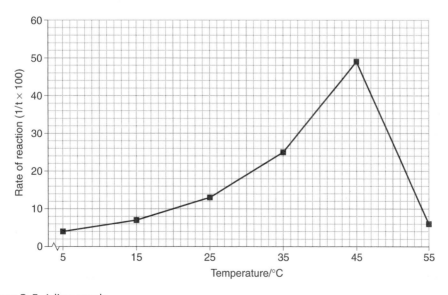

Figure 3.5 A line graph

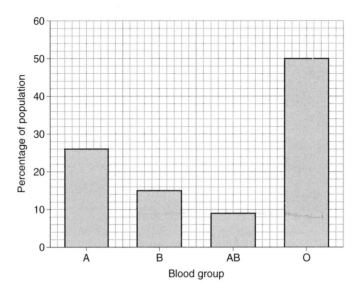

Figure 3.6 A bar chart

★ **Histogram:** used to show the frequency distribution of a continuous variable (one which can take any value within a given range), e.g. individual milk yield per cow (Figure 3.7).

★ **Scattergraph:** used to show the relationship between individual data values for two interdependent variables, e.g. height and weight of students (Figure 3.8).

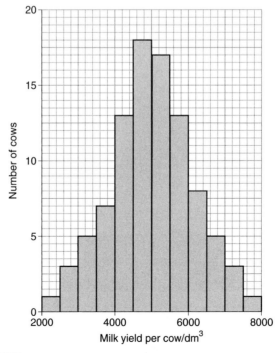

Figure 3.7 A histogram

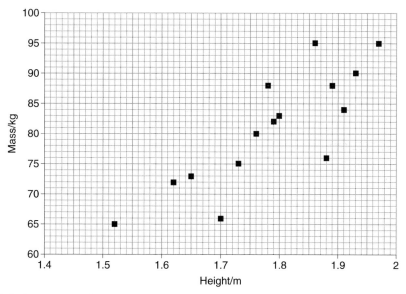

Figure 3.8 A scattergraph

You may also be required to draw other pictorial representations of data, such as a pie-chart, or (in ecology) kite diagrams or pyramids of number/biomass/energy.

Any graphs, bar charts or histograms should be plotted on graph paper. The horizontal x axis is usually used for the independent variable (manipulated by the experimenter, e.g. temperature) and the vertical y axis is used for the dependent variable (which is measured, e.g. enzyme activity). In a scattergraph, where the variables are interdependent, the axes may be orientated either way round. The following mnemonic (memory aid) may help you remember the key features of a graph.

S   **S**cale should be appropriate so that: the graph fits on the paper and is large enough to be seen easily; the points can be plotted accurately; values can be easily and accurately read from the graph.

P   **P**lotting should be accurate.

A   **A**xes should be the correct way round and labelled with units.

C   **C**urve should be appropriate: a smooth curve; or straight, ruled lines joining successive points; or a line of best fit.

K   **K**ey should be used to identify the curves if more than one set of data are plotted.

Extrapolation, or extension, of the curve beyond the range of your observations is usually unwise. When plotting and drawing curves from the data supplied, only those values given should be plotted as points. Many students make the mistake of extrapolating curves to zero, especially if the data relate to time or temperature.

An example graph for the data given in Table 3.2 is plotted in Figure 3.9.

Temperature/°C	Rate of reaction (1/t × 100)
5	2.1
15	4
25	6.9
35	13.3
45	20.8
55	6.8

Table 3.2

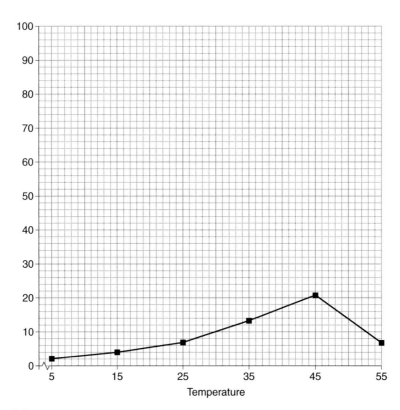

Figure 3.9

See if you can identify all the mistakes made in this graph and then draw it as it should have been.

A suggested answer is provided in Chapter 7, Making Connections: answers to exercises, on pages 76–83.

Using data

All synoptic questions containing experimental data will require you to *use* the data in some way. This might start with your having to extract one or more pieces of information from a table or graph, and it is vital that you do so accurately. The most common mistakes when taking readings from graphs (apart from accuracy) are: using the wrong axis, e.g. giving the temperature at which enzyme activity is at a maximum, rather than the actual figure for enzyme activity; using the wrong line if there is more than one line graph on the same axes; or failing to convert the figure from the graph into appropriate units, e.g. when dealing with very large numbers such as in bacterial growth curves. This last point is particularly important as there is often a mark for giving the correct units.

There are several different types of questions relating to data, which can be identified by their injunction (opening word).

DESCRIBE: this asks you to describe the data in the table or graph. A description of the trends and patterns shown by results does *not* need to include reference to every measurement or observation. It should be a summary, although you should identify all the significant points of change. If there are strange results that do not fit into the general pattern (anomalous results), then you should mention these in your account. It is important to be as specific as possible about the trends and patterns in the data, so try to use figures, e.g. 'the rate of activity increased between 5 minutes and 25 minutes', or (even better) 'the rate of activity increased by 18 arbitrary units s^{-1} (or 30%) between 5 minutes and 25 minutes'. Do *not* use expressions such as 'at the start/end', or 'faster/slower' if there is no time reference.

Question 1 is an example of a 'describe' question.

Question 1

An experiment was carried out to investigate the changes in concentration of glucose in the blood of an athlete during a marathon race. Measurements were made of the concentration of glucose at the beginning of the race and at 30 minute intervals during the race, which lasted 150 minutes. The results are shown in the table below.

Time during race / min	Concentration of glucose/μg mm^{-3}
0	60
30	26
60	24
90	23
120	22
150	22

Describe the changes that take place in the concentration of glucose in the blood throughout the race.

(2 marks)

In this case, a good answer to the question above would be as follows:

> The concentration of glucose drops quite rapidly from 60µg mm⁻³ to 26µg mm⁻³ in the first 30 minutes and then remains more or less constant (only dropping by 4µg mm⁻³) for the next 120 minutes.

Remember that 'describe' means 'give an accurate account of the main points' and an *explanation* of the data is *not* necessary.

CALCULATE: this requires you to carry out a calculation (perhaps having taken readings from a table or graph), showing your working and providing the appropriate units. Many students encounter problems when attempting to calculate *rates of reaction* or *percentage (%) changes*. Remember that:

$$\text{rate (of reaction)} = \frac{\text{quantity of product formed}}{\text{time taken}}$$

$$\text{percentage change} = \frac{\text{increase/decrease in quantity} \times 100}{\text{original quantity}}$$

Question 2 is an example of a 'calculate' question.

Question 2

An experiment was carried out to investigate the effect of different concentrations of caffeine on the reaction times of 20 human volunteers. Groups of 4 participants were given a drink containing either 0 (control), 1, 2, 5 or 10 arbitrary units of caffeine. After 20 minutes, their reaction times were tested and the results are shown in the table below.

Caffeine concentration/arbitrary units	Mean reaction time/seconds
0 (control)	0.56
1	0.44
2	0.40
5	0.38
10	0.37

Calculate the percentage decrease in reaction time between the control and the drink containing 10 arbitrary units of caffeine. Show your working.

(2 marks)

In this case, a good answer to the question above would be as follows:

> Decrease in reaction time = 0.56 – 0.37 = 0.19 seconds. Therefore, the percentage decrease in reaction time = (0.19/0.56) × 100 = 33.9%.

EXPLAIN: this requires you to give reasons for the trends and patterns in data, with reference to biological facts or concepts, using appropriate terminology. In this case, a description is *not* required.

Question 3 is an example of an 'explain' question.

Question 3

Amyloglucosidase is an enzyme which breaks down starch to glucose, by removing glucose units stepwise from starch molecules. The effect of temperature on the activity of a commercial preparation of amyloglucosidase is shown in the graph below.

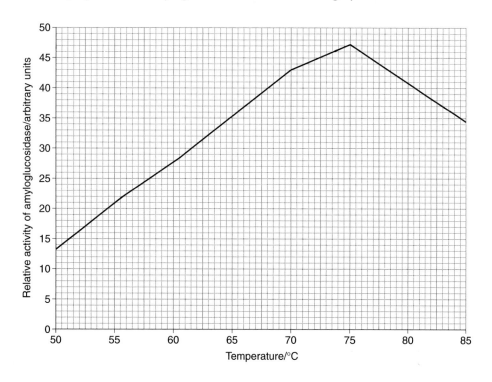

Explain why temperature affects enzyme activity as shown in the graph.   *(4 marks)*

In this case, a good answer to the question above would be as follows:

> Enzyme activity increases with increasing temperature (between 50°C and 75°C) due to the greater kinetic energy of the molecules in the system. This means that there will be more successful collisions between enzyme and substrate, forming more enzyme-substrate complexes and so more product. However, the bonds holding the enzyme active site in a particular shape will also vibrate more as the temperature rises. Above 75°C, some of these bonds start to break, distorting the shape of the active site. This means that fewer enzyme-substrate complexes are formed and enzyme activity decreases.

COMPARE: this requires you to point out similarities *and* differences between two sets of data (A and B). Remember that you must compare A with B according to each criterion you use, e.g. 'Both A and B showed the fastest rate of reaction in the first 5 minutes', or 'The reaction ceased after 15 minutes in A, but continued until 40 minutes in B'.

Question 4 is an example of a 'compare' question.

Question 4

An investigation was carried out to study the lead content of grasses growing by the side of a road. Samples of grass shoots were collected from an area alongside a road in a large city in Britain. The samples were taken each month from May 1989 to May 1990 and the lead content (μg of lead per g dry mass) of the grass shoots was determined. The data are shown in the graph below.

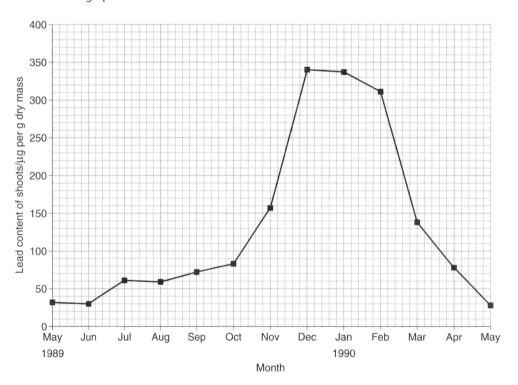

Compare the changes in lead content of the grass shoots between May 1989 and December 1989 with the changes between January 1990 and May 1990.   *(3 marks)*

In this case, a good answer to the question above would be as follows:

The lead content of the grass shoots rises by 308μg (from 32μg to 340μg) between May and December 1989, but falls by 309μg (from 337μg to 28μg) between January and May 1990. Therefore, the increase in 1989 is very nearly identical to the decrease in 1990. The rate of increase in lead content in 1989 is slower than the decrease in 1990.

Finally, there are fluctuations in the increase in 1989 (with two small decreases observed in June and August), but the decrease in 1990 is constant (although not even).

SUGGEST: this requires you to use your biological knowledge to put forward an answer in an unfamiliar situation. There is often not one 'right' answer, but any reasonable (biologically correct) suggestion could earn full marks.

Question 5 is an example of a 'suggest' question.

Question 5

The bar chart below shows the total light reaching the floor in two areas of a temperate deciduous wood in Europe over a period of one year. One area had recently been coppiced and the other area had never been coppiced (non-coppiced).

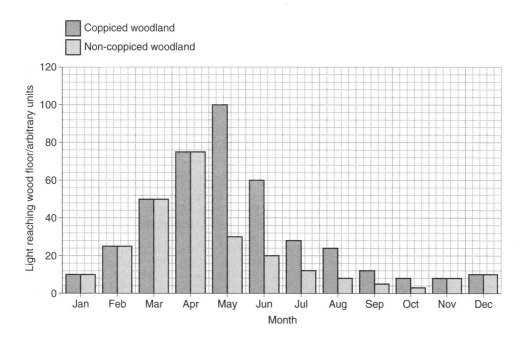

Suggest how the differences in light levels in the coppiced and non-coppiced areas affect the growth of plants on the floor of the wood. *(2 marks)*

In this case, a good answer to the question above would be as follows:

The greater light levels (May to October inclusive) in the coppiced areas would mean that plants on the floor of the wood could photosynthesise more, leading to more growth in these areas.

EXERCISE 4

Use the advice given in this section on 'using data' to answer the following question.

An investigation was carried out into the water balance of the body in different situations. A record was kept of the water intake and water output of a student on three separate days: a normal day spent indoors in warm temperatures, a day spent outside in cold weather and a day during which the student ran a cross-country race. On each day, the total water uptake by the body was the same as the water output.

The results are shown in the table below.

Water source	Normal day	Cold day	Day of race
Intake /cm^3			
Drinks	1500	1500	(i)..........
Food	800	800	1000
Oxidation of food	300	300	500
Total	2600	2600	(ii)..........
Output /cm^3			
Urine	1100	1500	1200
Sweat	1000	600	3000
From lungs	400	400	600
In faeces	100	100	100
Total	2600	2600	(iii)..........

a) Complete the table by calculating the volume of water intake from drinks and the total water intake and output for the day of the cross-country race. Write your answers on the lines labelled (i), (ii) and (iii) in the table. *(2)*

(b) (i) Compare the water intake and output on a normal day with that on a cold day *(3)*

(ii) Describe and explain the differences in water output on the normal day and on the day of the race. *(3)*

(c) (i) Explain how the oxidation of food produces water in the body. *(2)*

(ii) Suggest why more water is produced by oxidation of food on the day of the race. *(2)*

(d) During the race, the secretion of adrenaline increases. Describe and explain *one* physiological effect that this increase in adrenaline would have on the student. *(2)*

(Total 14 marks)

Edexcel

A suggested answer is provided in Chapter 7, Making Connections: answers to exercises, on pages 76–83.

Text comprehension

Synoptic questions in examinations may contain a short section of text on a particular topic, which acts as stimulus material for the question. The topic will usually be synoptic in nature, i.e. it will include information from more than one unit. Applying your biological skills to analyse this text is also synoptic. In order to score well on text comprehension questions, it will be useful to remember the following points.

★ Read the whole passage through quickly, highlighting or underlining key facts or concepts.

★ If you are asked to define or explain any biological terms in the passage, you should do so in context, i.e. explain the terms *as they are used in the passage.*

★ Remember that some of the information needed to answer questions will be included in the passage (look here first), but some will have to come from your own biological knowledge, so you should be prepared to use what you know.

The following passage is considered synoptic as it addresses ideas on evolution together with an appreciation of predator–prey relationships. Read through the passage and then consider the associated questions and suggested answers.

Predator-prey arms races

Species that use camouflage evolve to resemble their background. The behavioural strategy is to be immobile and try to be invisible. Of course, many predators detect prey using their sense of smell. In this case, animals may be camouflaged by being either odourless or similar in odour to their background. It is important to remember that an animal is only camouflaged if it settles in the right place. The European grasshopper exists in both green and yellow forms. In nature, the green forms tend to live in green habitats and the yellow form in yellow habitats. Similarly, studies on the peppered moth have shown that prey which is hard for predators to detect survive longer than those which are easy to find.

It should be remembered that predators will also have an advantage if they are camouflaged. The spotted coat of leopards, for example, makes them difficult to see on low tree branches because they blend in with light patterns created by sunlight penetrating the foliage. Some predators also have markings that disguise them as a harmless species, enabling them to get close to their prey. One such example is the blennid fish, which looks and behaves like a harmless cleaner wrasse, but which will attack if the opportunity arises.

a) (i) Explain how a species of prey may become progressively better camouflaged in its natural habitat.

> Within a population, there will be variation in the effectiveness of camouflage between individual organisms. This means that organisms which are less well camouflaged are more likely to be seen (and so eaten) by predators. If camouflage is genetically controlled, at least to some extent, reproduction between the surviving organisms will lead to the next generation being better camouflaged. Selection will continue in this next generation, again picking those organisms which are better camouflaged. This selection pressure will lead to the species of prey becoming progressively better camouflaged in its natural habitat.

(ii) Suggest what effect this improvement in camouflage will have upon animals which feed on this species of prey.

> As the prey become better camouflaged, it will be increasingly difficult for predators to find and consume them. This means that there will be selection pressure on the predators for improved visual abilities. Predators that are better at spotting prey are more likely to survive and reproduce, leading to selection for better eyesight (or sense of smell for predators using this method).

b) The blennid fish is an example of *aggressive mimicry*, in which a predator evolves to resemble a harmless species in order to get close to (and attack) prey. Suggest how prey organisms could also use mimicry.

> Prey organisms that are edible and/or harmless may mimic inedible or dangerous species in order to deter predators from attack. For example, hover-flies have no sting, but they may be avoided by predators because they look like wasps.

c) Suggest **one** *disadvantage* that there might be for organisms being well-camouflaged.

> Sometimes organisms may wish to be seen. For example, when defending a territory or seeking a mate.

d) The co-evolution of predators and prey is sometimes referred to as an *arms race*. Explain what is meant by this term.

> Improvements in the ability of predators to detect prey will lead to selection pressure on the prey organisms to develop better camouflage. This, in turn, will lead to selection pressure on predators to develop even better detection mechanisms. Therefore, each improvement in the ability of predators to catch prey will lead to an improvement in prey to avoid capture. This constant cycle of adaptation and counter-adaptation is known as an arms race.

EXERCISE 5

Read through the passage and then answer the questions which follow it.

> DNA replication is the process by which a DNA molecule produces two exact copies of itself. This process is controlled by the enzyme DNA polymerase. Hydrogen bonds break between the two polynucleotide chains and the parent DNA molecule unwinds. Each DNA strand then acts as a template for the synthesis of a new strand. Free nucleotides line up opposite their complementary bases and are joined together by condensation reactions to form two new DNA molecules. This process is known as semiconservative replication, as each new molecule contains half the original parent DNA molecule. In other words, the two daughter strands of DNA each consist of one strand of newly synthesised DNA and one strand of parental DNA. It is a common mistake to think that DNA replication is the same as transcription, which takes place during protein synthesis. There are several important differences between these two processes and it is vital that they should not be confused.

a) Explain what is meant by the following terms used in the passage.
   (i)   *hydrogen bond*   (2)
   (ii) *nucleotide*   (2)
   (iii) *condensation reaction.*   (2)

b) A sample of 10 μg of DNA underwent three successive divisions. Calculate:
   (i)   the total mass of DNA present after the three divisions   (2)
   (ii) the percentage of parental DNA in this final sample.   (2)

c) State **two** differences between DNA replication and transcription.   (2)

(Total 12 marks)

A suggested answer is provided in Chapter 7, Making Connections: answers to exercises, on pages 76–83.

Essays

Essays are ideal for testing synoptic skills because they require you to: bring together principles and concepts from different areas of biology and apply them in a particular context; and to express your ideas clearly and logically, using appropriate specialist vocabulary. Synoptic essays test your ability to *describe* and *explain* biological processes, and also examine your understanding of biological *principles* and *concepts*. You will be expected to develop an *argument* and present evidence for and against a statement. You will also be expected to show that you are aware of the *implications* and *applications* of modern biology. Marks are awarded for the *style* of presentation, the *selection* of appropriate material and the quality of written *communication*.

Tips

★ Read the question(s) carefully: it is surprising how many candidates misread questions, for example mixing up the terms 'mitosis' and 'meiosis', or 'deforestation' and 'desertification'.

★ Choose the essay carefully: if you have a choice of titles, you should think about which will earn you most marks. It is worth spending a short period of time jotting down what you know about a topic in order to make the right choice. There is nothing worse that writing half an essay on one topic and then abandoning it to start on the question you should have picked in the first place!

★ *Before* you start writing the essay you need to jot down all of your ideas (see 'Brainstorming' on page 9). Then organise these ideas into a logical sequence, ignoring issues that you decide are not relevant. You can probably group different ideas together, which means that you are starting to get the sub-sections or paragraphs for your essay. This forms your essay plan.

★ Review your plan to make sure that it is *balanced*, i.e. that it contains a range of ideas from several parts of the specification (AS *and* A2). For example, if the essay is about 'nitrogenous excretion in living organisms', make sure that you include a wide range of organisms and not just mammals (or even worse, just humans). Similarly, if the question specifies animals *and* plants, make sure that you give approximately equal weighting to both – don't write an essay on animals and then add on a brief paragraph about plants at the end.

★ Make sure that the essay is *structured*. It should start with an introductory paragraph which 'sets the scene' and introduces the topic in general terms. This should be followed by the main body of the essay, where you present your detailed discussion or argument. The essay should finish with a concluding paragraph which summarises the overall content. While you are writing, it is a good idea to refer back to your plan and tick-off points that you have covered. This ensures that you do not repeat yourself or miss anything out. It also helps you to avoid going 'off the point' and writing an essay about an entirely different topic!

★ Diagrams can be useful in an essay, but if you are going to use diagrams it is important to remember the following points: diagrams should be large, clearly drawn and accurately labelled or annotated; do *not* repeat information in a diagram that is already covered by what you have written – diagrams should complement the writing and add new information; be careful of time if including several diagrams – they often take longer than a written paragraph that would earn just as many marks. In general, if you are not sure whether to include a diagram in your essay, the best option is probably to leave it out.

★ Don't forget that you will also be judged on your *communication skills*. You should right your answer in 'free-prose', i.e. in sentences and paragraphs (*not* bullet points). Try to make sure that your writing is well-organised and legible, with good grammar and spelling.

The only effective way to develop your essay-writing skills is to *practise* as much as possible. The following system should help you to develop your skills in order that you can write consistently good essays under examination conditions.

Step 1 For a given essay title, complete the essay plan template below. It may help to discuss your ideas with your teachers and fellow students. Eventually, you will become skilled at planning essays in this way.

ESSAY PLAN TEMPLATE

Title:

Introduction:

Paragraph 1:
Paragraph 2:
Paragraph 3:
Paragraph 4:

Conclusion:

An example of a specific essay plan template is given on the next page.

EXERCISE 6

Produce an essay plan template for the title 'ATP and its roles in living organisms'.

A suggested answer is provided in Chapter 7, Making Connections: answers to exercises, on pages 76–83.

ESSAY PLAN TEMPLATE

Title: Chemical coordination in animals and plants

Introduction: The need for chemical coordination; general principles of chemical coordination.

Paragraph 1: Endocrine control in animals – nature of hormones; glands; principles of hormone action.

Paragraph 2: Animal physiology - sexual reproduction; control of blood glucose; osmoregulation.

Paragraph 3: Plant growth substances (auxins, giberellic acid, ethene, cytokinins, abscisic acid).

Paragraph 4: Plant physiology – growth, seed dormancy, leaf fall, root growth, bud development.

Conclusion: A large number of substances interact to coordinate physiology and behaviour in plants and animals.

Step 2 Practise completing essay plans under examination conditions, taking no more than 10 minutes. Discuss these plans with your teachers and fellow students, and then write the essays in your own time.

Step 3 Having practised producing plans under examination conditions, you should now practise writing essays in these conditions. Make sure that you only spend as long writing the essay as you will in the actual examination.

Step 4 You should now feel confident to practise writing unseen essays under examination conditions. A selection of possible titles is provided below.

★ The central role of DNA in living organisms.

★ The roles of pigments in living organisms.

★ Gas exchange in flowering plants and mammals.

★ The functions of proteins in plants and animals.

★ Genetic variation and speciation.

★ The chemical and biological control of insect pests.

★ The roles of enzymes in the control of metabolic pathways.

★ The role of carbohydrates in living organisms.

★ Energy flow through ecosystems.

★ Homeostasis and the principles of negative feedback.

Finally, it would be very useful for you to *mark* some essays. If you understand how the examiners mark synoptic essays in your examinations, you should be able to write essays that earn high marks. Further advice on writing and marking essays is given in Chapter 5, Analysing Synoptic Questions on page 35, but first try Exercise 7 to get an idea of what you should be looking for when marking (and therefore writing!) an essay.

EXERCISE 7

The essay below was written by a student under examination conditions. Read the essay carefully and then decide what mark to award, based on the following criteria:

- **Scientific content (S):** marks range from 0 (no correct information) to 12 (an excellent essay demonstrating a sound understanding of the topic and drawing material from two or more units of the specification).

- **Balance (B):** marks range from 0 (very limited account with much irrelevance and many errors) to 2 (a balanced essay covering all of the main areas relevant to the title with few, if any, errors).

- **Coherence (C):** marks range from 0 (essay style not used and/or very poor standard of spelling, punctuation and grammar) to 2 (introduction and conclusion included; logical structure; continuous prose throughout; and sound spelling, punctuation and grammar).

(Total marks (0-16): S + B + C)

The roles of water in the lives of organisms

Water is perhaps the most important molecule for the survival and life of organisms. On the surface of the planet there is obviously much more area of water than land, which shows its significance.

It is mostly due to its specific properties that water is so useful. Perhaps the most obvious is that it has a very high specific heat capacity. It is noticeable that water is a much more stable environment for organisms to live in, as it does not cool to rapidly or heats up like air. This means that organisms living in water do not have to keep re-adjust their body temperature for survival.

Water provides support for marine organisms such as jelly fish, both from the outside and inside. On the outside it allows for movement, as floating is possible, so joints are not required or hard skeleton for muscles to act upon. On the inside, it also provides support, so movement can be achieved by pushing water into the front, propagating itself along.

On earth water inside an organism also provides support. For example earthworms employ a hydrostatic skeleton, which allows muscles to be contracted against an incompressible substance.

Water is a universal solvent. It provides medium in which substances can be dissolved. It is present in cytoplasm of all organisms, so they cannot survive without it. Water is present in blood to allow digested substances such as glucose to be carried along to respiring cells and allows easier diffusion of materials between cells. Examples of this is in alveoli and cappillaries. It is a lubricant and a transport substance.

Water allows toxic materials, such as urea to be dissolved so it does not harm the organism that has produced it during metabolic activities.

Water acts as a shock absorber, for example in cavities in the brain (the serebrospinal fluid) which cushions the delicate organs against damage. It is present in joints to allow easy movement and protection against friction. In the eye to give it shape and support (as aqueous and vitreous humour). It is present in mucus in the mouth to allow easy swallowing of food, in the gut for moving the food along and dissolving enzymes to break down food and antibodies or bacteria for protection against disease.

Water is essential in homeostasis (in urine, as I have already mentioned and in control of the body temperature as specific heat capacity allows for cooling of the skin in sweat.

In plants water has significance importance. It allows for support, as water is an incompressible fluid. It fills the cells which are surrounded by cell walls and so keep them turgid.

It is also a transport medium, just like in animals) substances, such as ions are dissolved in it and carried in the xylem. The sugars produced by photosynthesis are also dissolved in water and carried to respiring cells.

Water is an essential ingredient in photosynthesis as it is where the oxygen comes from (Hill's reaction) as water is split into oxygen and hydrogen ions (and also electrons).

Water is important in moving substances up the transpiration stream. Water evaporates from the surface of the leaf, and therefore the substances from the soil are drawn up.

Another method of water movement up the xylem is by cohesion-tension theory. The molecules of water are attracted to the walls of the xylem (adhesion) due to its polar structure, and to each other (cohesion) so it moves up the stream because the walls of the lignified xylem can take the pressure.

Fertilisation in some organisms relies on water. Fish release their eggs into the river, so that the male can then fertilise them outside the body. Some plants rely on water to transport pollen in the water.

Terrestrial animals are also dependent on the lubricating property of water. Human release mucus by their sexual organs to allow for easier penetration. Sperm is also released in a semi-liquid (containing water) this allows for easier movement to reach the egg. The mucus produced by females in the vagina, also protects against infections as water dissolves the antibodies.

One other aspect of water I have not mentioned is its insulating properties. Water expands on heating, due to its original property and structure. The ice which forms, floats on the rivers, so heat does not escape from the water below and organisms are able to survive.

Water vapour in the atmosphere also acts as a greenhouse gas, allowing for radiation to warm up the earth's surface, but not allow the heat to escape back into the atmosphere.

From all this information it is clear how many functions water has in the living organisms. Water is not only present in eyes to support and keep its shape, but to lubricate it and protect it from infection and it also helps to keep our environment warm. It makes up about 65% of human mass and up to 95% in plants. This clearly indicates its great significance.

Edexcel

A suggested mark for this essay is provided in Chapter 7, Making Connections: answers to exercises, on pages 76–83.

Problem solving and applications of biology

All A-level Biology specifications aim to promote an awareness and appreciation of the significance of biology in 'personal, social, environmental, economic and technological contexts'. It is therefore very likely that you will find some synoptic questions which are designed to test *applications* of biology, or to *solve problems* in our everyday lives. Instead of providing data about plants grown in a research laboratory, the questions may refer to a crop plant, or to microorganisms used in yoghurt manufacture or brewing beer. These questions may also include environmental, health and even ethical issues, such as the use of genetically modified organisms, or *in vitro* fertilisation and genetic counselling in humans. You must remember that questions of this type are testing your understanding of the *biology* involved and your awareness of some of the arguments surrounding an issue, rather than expecting you to give your personal opinions.

Questions on problem-solving and applications of biology are often synoptic in nature because they require you to use different parts of the specification to answer questions and (more commonly) to use your knowledge to answer questions set in a novel situation. Some of these 'novel' situations will simply be adaptations of familiar procedures, e.g. the immobilisation of sucrase for the commercial production of fructose rather than the (more familiar) immobilisation of lactase to produce lactose-free milk. Other 'novel' situations will be genuinely novel and will require you to bring together principles and concepts from different areas of biology and apply them to a particular problem. Further advice on tackling this type of question is given in Chapter 5, Analysing Synoptic Questions on page 35.

CHAPTER FOUR

HOW ARE SYNOPTIC SKILLS ASSESSED?

The manner in which synoptic skills are assessed is slightly different for each awarding body (examination board). Therefore this section is divided up into four parts, one for each of the major awarding bodies (AQA(A), AQA(B), Edexcel and OCR), considering *where* the skills are assessed and *how* they are assessed. It is important to remember that synoptic skills account for 20% of the marks for *all* awarding bodies and that the question styles used are broadly similar. Synoptic questions require you to make connections between at least two units and to use your skills and ideas in new contexts. Structured question styles include: data analysis; text comprehension; problem solving; and applications of biology. The only other type of synoptic question is the essay.

AQA (A)

Unit	Time	Synoptic question style(s)	Percentage of synoptic marks	Percentage of A-level marks
6 or 7	$1\frac{1}{2}$ hours	Various structured questions (*)	5%	15%
8a or 9a	$1\frac{3}{4}$ hours	Two structured questions and one essay (chosen from two titles)	10%	10%
8b or 9b	N/A	Coursework (*)	5%	10%

[*Note:* N/A means not applicable; (*) indicates a mixture of synoptic and non-synoptic marks within that unit; Units 6 and 8 lead to a qualification in Biology and Units 7 and 9 lead to a qualification in Human Biology.]

AQA (B)

Unit	Time	Synoptic question style(s)	Percentage of synoptic marks	Percentage of A-level marks
5a	$1\frac{1}{4}$ hours	Various structured questions (*)	3.5%	7.5%
5b	N/A	Coursework (*)	2.5%	7.5%
6 or 7 or 8	2 hours	Several structured questions and one essay (chosen from two titles) (*)	14%	20%

[*Note:* N/A means not applicable; (*) indicates a mixture of synoptic and non-synoptic marks within that unit; Units 6, 7 and 8 refer to different options – 'Applied Ecology', 'Microbes and Biotechnology', or 'Behaviour and Population'.]

Edexcel

Unit	Time	Synoptic question style(s)	Percentage of synoptic marks	Percentage of A-level marks
5B or 5H	$1\frac{1}{2}$ hours	Various structured questions plus a free-prose (*)	11%	16.7%
6 (#)	1 hour 10 mins	Two structured questions and one essay (chosen from two titles)	9%	9%

[*Note:* (*) indicates a mixture of synoptic and non-synoptic marks within that unit; 5B refers to the Biology qualification and 5H refers to the Human Biology qualification; (#) indicates that Unit 6 also contains coursework (non-synoptic); see page 5 for more details about free-prose questions.]

OCR

Unit	Time	Synoptic question style(s)	Percentage of synoptic marks	Percentage of A-level marks
2805 [01–05]	$1\frac{1}{2}$ hours	Various structured questions (*)	5%	15%
2806 [01]	$1\frac{1}{4}$ hours	Several structured questions plus one essay (compulsory)	10%	10%
2806 [02 or 03]	N/A [02] $1\frac{1}{2}$ hours [03]	Coursework (Paper 02) or practical examination (Paper 03) (*)	5%	10%

[*Note:* N/A means not applicable; (*) indicates a mixture of synoptic and non-synoptic marks within that unit; Unit 2805 has 5 options [01–05], of which candidates choose one.]

CHAPTER FIVE

ANALYSING SYNOPTIC QUESTIONS

This section of the book considers the five main types of synoptic question: data analysis, text comprehension, commercial applications of biology, problem-solving and essays. Each question type is addressed in turn, with advice on how to tackle the question in order to score highly. Additional questions are then supplied (in the form of exercises), so that you can practise your skills. Once you have completed this chapter it would be useful to attempt the mock examinations on pages 60–75, preferably under examination conditions.

Data analysis

Question 1

In an investigation on temperature regulation in humans, a healthy volunteer was immersed in a bath of water at 10°C. During the investigation he kept one hand in the air on the side of the bath, and one hand submerged in the water. His core body temperature, the temperature of the hand in the air, and the temperature of the hand in the water were recorded immediately after immersion and then every minute for seven minutes. The air temperature in the room remained at 26°C during the investigation.

The results are shown in the table below.

Time / mins	Core temperature / °C	Temperature of hand in air / °C	Temperature of hand in water / °C
0 (start)	37.1	25.0	23.2
1	37.1	25.0	18.0
2	37.1	25.0	16.4
3	37.1	25.0	14.8
4	37.1	25.0	14.0
5	37.2	25.2	13.2
6	37.2	25.0	12.8
7	37.2	25.0	12.4

a) Plot a suitable graph on graph paper to show how the temperature of the hand in air and the temperature of the hand in water changed during the course of this investigation. *(5)*

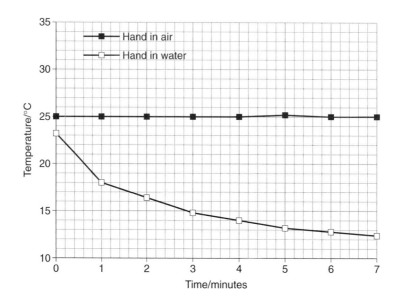

b) Compare the changes in temperature of the hand immersed in water with the temperature of the hand in the air. *(2)*

> The temperature of the hand in air stays almost constant, but the temperature of the hand in water decreases steadily. The temperature of the hand in air is always higher than that of the hand in water (1.8°C higher at the start and 12.6°C higher after 7 minutes).

c) Suggest reasons for the differences you have described in (b). *(2)*

> The hand in air stays more or less at room temperature (26°C), whereas the hand in water cools towards the water temperature (10°C). The hand in water loses heat by conduction and will also have a reduced blood flow due to vasoconstriction.

d) Explain why the core temperature remained almost constant during this investigation. *(3)*

> Vasoconstriction results in decreased blood flow to the skin, which reduces heat loss. The metabolic rate increases (and there may be shivering) to generate heat and keep the core temperature constant.

e) The diagram below shows part of the mechanism that regulates body temperature in humans.

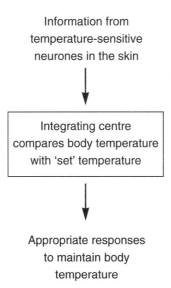

Information from
temperature-sensitive
neurones in the skin

↓

Integrating centre
compares body temperature
with 'set' temperature

↓

Appropriate responses
to maintain body
temperature

(i) State where in the body the 'integrating centre' is situated. *(1)*

> In the hypothalamus.

(ii) Describe how information from temperature-sensitive neurones in the skin is conveyed to the integrating centre. *(2)*

> Nerve impulses are passed from the receptors along sensory neurones to the integrating centre. The frequency of nerve impulses will vary according to the temperature of the skin.

(Total 15 marks)

Edexcel

EXERCISE 8

Bread dough is made by mixing together wheat flour (which contains starch), water and yeast. The dough rises because carbon dioxide produced by respiring yeast is trapped in the dough.

An investigation was carried out into the effect of adding glucose to bread dough. 20 cm^3 of dough, without added glucose, was kept at a temperature of 20°C. The volume was measured every ten minutes for one hour.

The experiment was repeated with a second sample of 20 cm^3 of dough to which 1 g of glucose had been added.

The results are shown in Table 5.1

Time / min	Volume of dough / cm^3	
	Dough without added glucose	Dough with added glucose
0	20	20
10	27	32
20	31	41
30	45	51
40	50	60
50	51	63
60	51	65

Table 5.1

a) (i)   Plot the data in suitable graphical form on graph paper.   *(5)*
    (ii) During the first 15 minutes, the mean rate of change in volume of the dough without added glucose was 36 cm^3 h^{-1}. Calculate the difference between this rate and the mean rate of change in volume of the dough with glucose over the same period. Show your working.   *(3)*
    (iii) What conclusions can be drawn from this investigation?   *(2)*
    (iv) Suggest an explanation for the effect of adding glucose to the bread dough.   *(2)*

b) In a further investigation, amylase was added to the bread dough without glucose. Suggest what effect this might have, giving reasons for your answer.   *(3)*

(Total 15 marks)

Edexcel

Question 2

An investigation was carried out into the excretion of urea in the urine of a healthy student on high and low protein diets.

The results for each diet are shown in the table opposite. The results are mean values for the second, third and fourth days after starting the diet.

Diet	Protein intake / g per day	Daily output of urea / g
High protein	189	21.4
Low protein	56	8.5

a) Calculate the percentage difference in the daily output of urea between the two diets. Show your working. *(2)*

> (12.9/21.4) x 100 = 60.3% or (12.9/8.5) X 100 = 151.8%

b) Give an explanation for this difference in the daily output of urea. *(3)*

> The low protein diet has fewer excess amino acids and therefore less deamination will be required. As a result, there will be less ammonia and therefore less urea produced.

c) Suggest why many aquatic organisms excrete ammonia, rather than urea, as the end product of protein metabolism. *(2)*

> Ammonia is more toxic than urea and therefore requires relatively large quantities of water for its removal. Aquatic organisms can release ammonia directly into the surrounding water and it will diffuse away. This saves the organisms from having to convert ammonia to urea, which is a process requiring energy.

(Total 7 marks)

Edexcel

EXERCISE 9

An investigation was carried out to compare the water potential of two tissues.

A series of sucrose solutions of molarities ranging from 0.05 to 0.40 mol dm^{-3} was prepared, each solution being placed in a separate beaker. The density of sucrose solution increases as the concentration of sucrose increases.

Half the 0.05 mol dm^{-3} solution (from beaker A) was transferred to another beaker (B). A small amount of a coloured substance was added to the solution in beaker B. This coloured sucrose solution was then used to half-fill three test-tubes. Into one test-tube was placed 5 g of sliced potato tuber tissue, into the second was placed 5 g of sliced beetroot tissue and the third test-tube was left as a control. The sliced tissues were carefully blotted before being placed in the test-tubes.

Synoptic Skills in Advanced Biology

The tissues were left in the test-tubes for 10 minutes. A syringe was then used to remove a drop from each test-tube and inject the drop below the surface of the solution in beaker A. Each drop was observed to see if it rose upwards, stayed suspended or sank to the bottom of the beaker.

Figure 5.1 illustrates the procedure used.

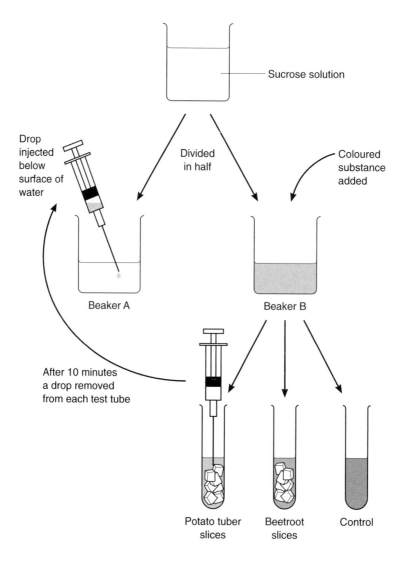

Sucrose solution

Drop injected below surface of water

Divided in half

Coloured substance added

Beaker A

Beaker B

After 10 minutes a drop removed from each test tube

Potato tuber slices

Beetroot slices

Control

Figure 5.1

The procedure was repeated for each of the sucrose solutions and the results recorded in the Table 5.2

	Concentration of sucrose solution / mol dm^{-3}							
Tube content	0.05	0.10	0.15	0.20	0.25	0.30	0.35	0.40
Potato tuber tissue	–	–	0	+	+	+	+	+
Beetroot tissue	–	–	–	–	0	+	+	+
Control	0	0	0	0	0	0	0	0

KEY

+ indicates that the drop rose upwards
– indicates that the drop sank to the bottom
0 indicates that the drop stayed suspended

Table 5.2

a) (i) Why was the coloured substance added to beaker B? *(1)*
   (ii) What was the purpose of the control tubes? *(1)*
   (iii) Why was the tissue blotted before it was added to the test-tubes? *(1)*

b) Explain why, in this investigation, some of the drops sank and why some rose upwards. *(4)*

c) State the conclusions that can be drawn from this investigation. *(2)*

d) Investigation of the cell contents of the two tissues showed that carbohydrate is stored in potato tuber cells as starch and in beetroot cells as sucrose. Suggest why this might affect the water potential of each tissue. *(2)*

e) Suggest **three** precautions that would need to be taken to ensure the reliability of these results. *(3)*

(Total 14 marks)
Edexcel

Question 3

Oestrogen compounds are often present in aquatic environments and may be concentrated near sewage outfalls and in water draining from landfill sites.

An investigation was carried out to determine the effect of different concentrations of an oestrogen compound (ethyl oestradiol) on the heart rate of *Daphnia*, a freshwater crustacean. The heart rate of each *Daphnia* was determined when it was immersed in a range of concentrations of the oestrogen compound.

Batches of five *Daphnia* were used. After a period of ten minutes in each concentration of the oestrogen compound, the heart rate was measured for three one-minute periods and the mean rate was determined.

The results are shown in the graph on the next page.

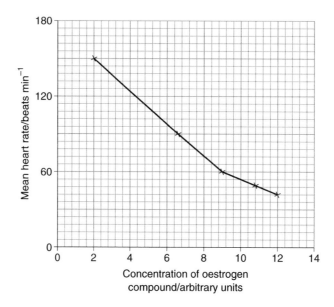

a) Explain why the *Daphnia* were left in each concentration of the oestrogen compound for ten minutes before the heart rate was determined. *(2)*

> To become acclimatised to the conditions, so that the heart rate measured
> was due to the concentration being studied, i.e. so that the oestrogen had
> time to cause an effect.

b) Explain why batches of five *Daphnia* were used and why the heart rate was determined three times at each concentration. *(2)*

> To obtain a mean value at each concentration in order to make the
> results more reliable.

c) Explain why temperature should be kept constant during the investigation, giving a reason for your answer. *(2)*

> Temperature will affect the metabolic rate of the Daphnia (due to its
> effect on enzyme activity), which will alter the rate of heartbeat.

d) What conclusions can be drawn from the results of this investigation? *(2)*

> That there is an inverse relationship between the concentration of ethyl oestradiol and heart rate, i.e. increasing the concentration of the oestrogenic compound decreased the rate of heartbeat.

e) Suggest how the changes in the rate of heartbeat caused by oestrogen compounds could affect the *Daphnia* in their natural surroundings. *(2)*

> It could cause increased mortality in Daphnia, either directly or by making them move more slowly (and so more are caught and eaten).

f) Suggest **one** further effect that the oestrogen compounds might have on the *Daphnia* and explain how this could affect the ecosystems in which they occur. *(2)*

> It could affect fertility in Daphnia, leading to an increase or decrease in population numbers.

g) A further experiment was carried out to investigate the effect of the oestrogen compounds on the rate of photosynthesis in the freshwater pond weed, *Elodea*.

The results are shown in the graph below.

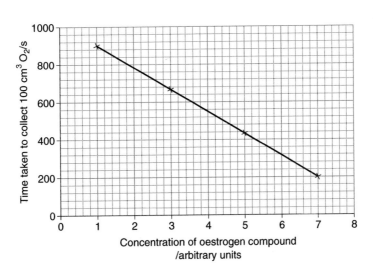

(i)   Describe the effect of increasing concentrations of the oestrogen compound on the photosynthetic activity of *Elodea*.   *(1)*

> *There is a direct relationship between the rate of photosynthesis of Elodea and the concentration of the oestrogen compound.*

(ii)   *Elodea* is a food source for herbivores. Explain how the presence of oestrogen compounds in the water of the ponds and lakes where *Elodea* occurs could affect the human food chain.   *(2)*

> *The oestrogen compounds would get passed on to herbivores and eventually accumulate in fish that are eaten by humans.*

(Total 15 marks)
Edexcel

EXERCISE 10

Figure 5.2 shows a water flea (*Daphnia*), an organism which is commonly found in freshwater habitats.

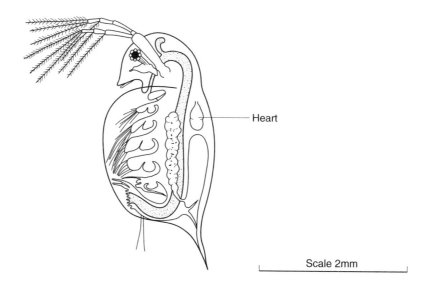

— Heart

Scale 2mm

Figure 5.2

a) (i)   Name the phylum to which *Daphnia* belongs.   *(1)*
     (ii)   Give **two** external features which are characteristic of this phylum.   *(2)*

b) An experiment was carried out to investigate the effect of different concentrations of caffeine on the heart rate of *Daphnia*. Specimens were kept in pond water, maintained at a constant temperature of 25°C. They were then placed individually on microscope slides and caffeine solution was added to each. After one minute, the heart rate of each *Daphnia* was determined. This procedure was repeated ten times for each concentration of caffeine.

The results are shown in Table 5.3

Concentration of caffeine / parts per million	Mean heart rate / beats per second
1000	5.6
100	4.9
10	4.4
1	4.1
0.1	4.0
0.01	3.9
0.001	3.8
0.0001	3.7
0.00001	3.7
0.0 (control)	3.7

Table 5.3

(i) Describe the relationship between heart rate and caffeine concentration. *(3)*

(ii) Calculate the percentage increase in heart rate between the control and the solution containing 1000 parts per million caffeine. Show your working. *(2)*

c) (i) What is the purpose of the control in this experiment? *(1)*

(ii) Suggest why the experiment was repeated at each concentration of caffeine. *(2)*

(iii) Explain why the specimens were kept at a constant temperature during the experiment. *(3)*

(Total 14 marks)
Edexcel

Text comprehension

Question 4

Read the passage below and then answer the questions which follow.

> Evolutionary change develops when a mutation occurs and survives the selective process (that is, when it is found to be either neutral or advantageous). For example, a GCT codon might mutate to GAT and we would obtain leucine instead of arginine in the protein.
>
> In about 20% of all mutations, because of the redundancy (degeneracy) of the code, a mutation might have no effect on protein structure; thus a mutation from GCT to GCA would affect only the DNA and might well have no functional effects – no matter what the third base in the GC codon, we always obtain arginine in the protein.
>
> The evolutionary process, then, involves a change (mutation) in the DNA which is incorporated into the ongoing gene pool of the evolving species and which can be reflected by a corresponding change in the amino acid sequence of the particular protein coded for by that gene.
>
> We might state as a basic rule that such a process will have to produce divergence when any two populations become isolated from one another, as the relative rarity of mutations and the finite size of populations make it statistically improbable that identical changes will be available for natural selection to incorporate into the gene pools.

a) Explain what is meant by each of the following terms.
   (i)   A GCT codon (paragraph 1)   *(1)*

> *A DNA triplet of nitrogenous bases, consisting of guanine, cytosine and thymine.*

   (ii) Redundancy (degeneracy) (paragraph 2)   *(1)*

> *There are more codons available than there are different types of amino acid.*

(iii) Gene pool (paragraph 3)   *(1)*

> The total number of alleles in a population.

(iv) Natural selection (paragraph 4)   *(1)*

> The means by which organisms best suited to the environmental conditions will survive.

b) (i) Suggest **two** ways in which 'mutation from GCT to GCA' (paragraph 2) might arise.   *(2)*

> Substitution of A for T, or an inversion mutation (if the next base along is A). This might be caused by ionising radiation or mutagenic chemicals.

(ii) Explain why such a mutation 'might well have no functional effects' (paragraph 2).   *(3)*

> Usually only the first two bases are necessary to code for the required amino acid, e.g. all codons starting with GC will code for arginine. Therefore the amino acid sequence of the protein produced might not be different. Furthermore, even if it was different, this change in sequence might not affect the structure or function of the protein.

c) (i) State **two** ways in which 'populations become isolated from one another' (paragraph 4).   *(2)*

> Geographical isolation occurs when populations become isolated due to a physical barrier, such as a river or mountain range.
> Behavioural isolation occurs when populations become isolated due to different behaviour patterns, e.g. incompatible courtship activities.

(ii) Outline the possible consequences of the 'divergence' (paragraph 4) which may result from such isolation   *(3)*

> Isolated populations will become progressively different from one another, due to changes in the gene pool. Eventually the two populations will be unable to interbreed to produce fertile offspring. At this point they can be regarded as being different species.

(Total 14 marks)
Edexcel

EXERCISE 11

Read the passage below about the effect of chilli peppers on the body, and then use the information in the passage and your own knowledge to answer the questions which follow.

Chilli peppers (*Capiscum frutescens*) contain a substance called capsaicin, which produces the painfully hot sensation that we receive when eating foods seasoned with chilli. It has been known for some time that capsaicin produces its effect on a particular type of sensory neurone. These neurones respond to painful stimuli such as high temperatures, or to inflammation of tissues. Exposure of the endings of these neurones to capsaicin allows calcium ions (Ca^{2+}) and sodium ions (Na^+) to flood into the neurone. This initiates an action potential that is carried along the neurone into the spinal cord and then to the brain, where it is interpreted as pain.

It was thought that the sensory neurones probably contain a protein in their cell surface membranes that acts as a receptor for capsaicin. This protein could work in a similar way to the receptor for acetycholine, which is found on the postsynaptic membrane at many synapses in the body. A group of researchers has recently identified the length of DNA which codes for this capsaicin receptor protein.

The researchers extracted mRNA of many different lengths from cells taken from dorsal root ganglia. The next step was to identify which of these lengths coded for the capsaicin protein. They mixed their mRNA samples with DNA nucleotides, and added an enzyme which made complementary lengths of DNA, called cDNA. They introduced samples of this cDNA into some cultured human cells, and allowed the cells to express the cDNA. They then exposed the cells to capsaicin. By identifying the cells that allowed calcium ions to flow into them when in the presence of capsaicin, the researchers were able to identify

> which of the many different lengths of cDNA coded for the capsaicin receptor protein. Once the cDNA had been identified, the amino acid sequence of the receptor protein could be worked out.
>
> The researchers produced a clone of cells containing this cDNA, which had the capsaicin receptor in their cell surface membranes. They found that the receptor responded not only to capsaicin but also to high temperatures, which we perceive as painfully hot. They also found that, if the cells were exposed to capsaicin for a long period of time, they became less sensitive not only to capsaicin but also to high temperatures.

a) Explain how capsaicin initiates an action potential in a sensory neurone (paragraph 1). *(3)*

b) Suggest why the researchers chose to use cells from dorsal root ganglia when they extracted mRNA (paragraph 3). *(2)*

c) Suggest why the researchers decided to try to identify the protein by extracting mRNA, rather than DNA, from these cells (paragraph 3). *(2)*

d) Explain what is meant by DNA nucleotides (paragraph 3). *(2)*

e) Name the enzyme that would be added to the mRNA and DNA nucleotides to make the cDNA (paragraph 3). *(1)*

f) Explain how the amino acid sequence of the capsaicin receptor could be worked out once the cDNA that coded for it was identified (paragraph 3). *(3)*

g) Capsaicin is often used to reduce pain in inflammatory conditions such as arthritis. Using the information in the passage, suggest how application of capsaicin to a painful joint might reduce the sensation of pain. *(2)*

(Total 15 marks)
Edexcel

Commercial applications of Biology

Question 5

An experiment was carried out to determine the effect of temperature on the activity of an enzyme digesting the protein gelatin.

Gelatin was incubated with the enzyme at a range of temperatures from 5°C to 60°C. The rate of amino acid production was measured over a three-hour period.

The results are given in the table below, expressed as rate of amino acid production in mg dm^{-3} h^{-1}.

Temperature / °C	Rate of production of amino acid / mg dm^{-3} h^{-1}
5	14
10	19
15	24
20	31
25	40
30	51
35	68
40	93
45	98
50	89
60	33

a) (i) Plot the data on graph paper. *(4)*

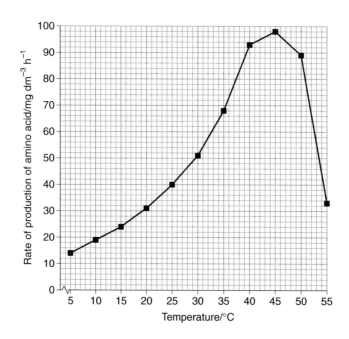

(ii) Comment on the effect of temperature on the activity of the enzyme as shown in the graph. *(3)*

> The rate increases steadily up to 40°C and then starts to slow, with an optimum rate at 45°C. This is due to an increase in the kinetic energy of the molecules in the system. More enzyme-substrate complexes are formed and so more product is made (per unit time). Above 45°C, there is a slow decline to 50°C and then a steep fall (from 89 mg dm^{-3} h^{-1} to 33 mg dm^{-3}h^{-1}) to 60°C. This is due to the higher temperatures breaking bonds within the enzyme, changing the shape of the active site, i.e. the enzyme is being denatured.

b) The experiment was continued at 45°C for a further 7 hours. At the end of this time, an additional 292 mg dm^{-3} of amino acid had accumulated.

(i) Calculate the mean rate of reaction during the 10 hours at 45°C. *(1)*

> $[(3 \times 98) + 292]/10 = 58.6$ mg dm^{-3}h^{-1}

(ii) Give **two** possible reasons for the difference between the rate at the end of 10 hours and the rate after 3 hours incubation. *(2)*

> The substrate might have been used up or the enzyme may have started to denature.

c) Protein-digesting enzymes can be used as an ingredient in biological washing powders.

(i) Suggest how the results of this experiment could be used to design a suitable washing programme using a biological washing powder. *(2)*

> The programme could be set at 40-45°C, with a washing cycle of 3 hours maximum.

(ii) Suggest possible advantages of using biological washing powders rather than non-biological detergents. *(2)*

> They can be used at lower temperatures and so save energy (and may be less damaging to fabrics). They will also be better at getting rid of biological stains, such as blood.

(Total 14 marks)

Edexcel

The toxicity of certain substances can be determined by the use of an LD_{50} test. LD_{50} is defined as the concentration of a substance which results in the death of 50% of a population of test organisms in a given time period.

An investigation was carried out to test the effectiveness of a new insecticide on the larvae of the mosquito, *Anopheles* sp. Groups of larvae were treated with the insecticide at concentrations ranging from 0.1 to 5.0 parts per million and the percentage mortality was calculated after 2 days. The results are shown as a graph in Figure 5.3.

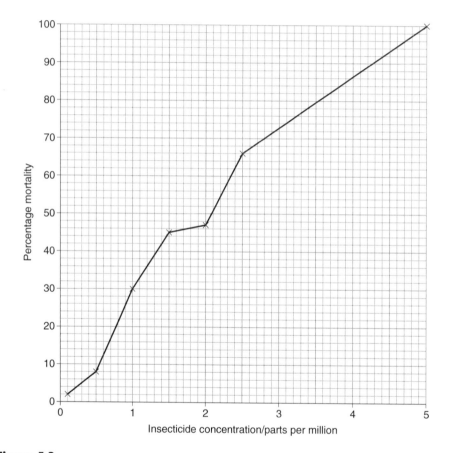

Figure 5.3

a) (i) From the graph, find the LD_{50} for this insecticide. *(1)*

   (ii) Suggest and explain **one** long-term effect on mosquito populations of repeated use of this insecticide at concentrations lower than the LD_{50}. *(2)*

b) Suggest a suitable control for this investigation and explain why it is necessary. *(2)*

c) The larvae of mosquitoes live in freshwater pools. Further tests on the insecticide showed that it was soluble in water and chemically stable. Suggest **three** reasons why the insecticide might be unsuitable for general use. *(3)*

d) Some insecticides function as inhibitors of the enzyme acetylcholinesterase, which breaks down acetylcholine. Suggest what effects this may have on the body of an insect. *(3)*

e) (i) State **two** advantages of the use of natural predators, rather than chemical insecticides, for the control of insect pests. *(2)*

   (ii) State **one** disadvantage of the use of natural predators, rather than chemical insecticides, for the control of insect pests. *(1)*

(Total 14 marks)
Edexcel

Problem solving in Biology

Question 6

A student noticed that the density of some plant species appeared to differ depending on how far the plants were from a main road.

The mean density (plants per m^2) of three plant species A, B and C was measured at different distances from the main road. The mean density of the same three plant species was also determined at the side of a narrower secondary road in the same locality.

The results of the investigations are shown in the diagrams below and on the following page.

Main Road

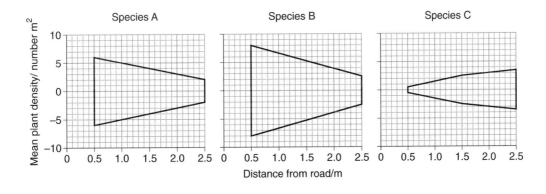

Secondary road

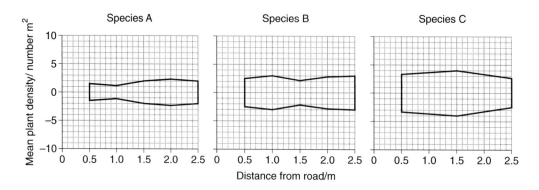

a) Describe a procedure the student could have used to determine the mean density of the three plant species. *(4)*

> Take several samples of a given area (using a quadrat) at each
> measured distance away from the road. Count the number of each of the
> plant species in each quadrat and calculate the mean. It would be
> important to carry out the investigation on a single day.

b) (i) Comment on the relationships between plant density and the distance from the main and secondary road for species A and B. *(4)*

> The mean density of species A and B decreases with distance from the
> main road, but varies very little with distance from the secondary road.
> The mean density of species A is lower than that of B, for both the main
> and secondary road areas. There is a lower overall mean density of both
> species near the secondary road than near the main road.

(ii) Comment on the ways in which the distribution of plant species C differs from that of plant species A. *(2)*

> The mean density of C inceases with distance from the main road and
> there is a greater density of C than A at the side of the secondary road.

c) In addition to determining the plant densities, the student measured the pH of soil samples taken at the same distances from each of the roads.

The results are shown in the graph below.

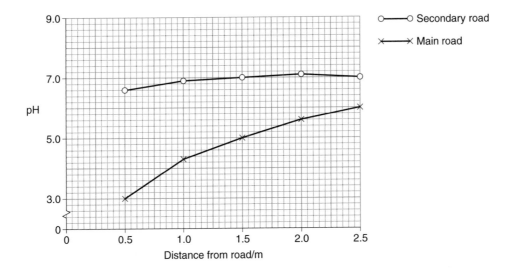

(i) Suggest an explanation for the differences between the pH of the soil at the side of the main road and the pH of the soil at the side of the secondary road. *(2)*

> There is more traffic on the main road and the acidic gases (carbon dioxide and nitrogen oxides) in exhaust fumes dissolve in rain and get washed into the soil. Therefore the soil pH is lower (more acidic) by the side of the main road, particularly at points close to the road.

(ii) Using the data given for pH, suggest an explanation for the distribution of the three species A, B and C. *(2)*

> A and B may be able to tolerate the low pH at the side of the main road (or grow better in acidic soils), but C cannot.

(iii) Suggest **one** factor, other than pH, which could account for the differences in density distribution of the plant species at the side of the main road. *(1)*

> Other toxic chemicals, such as lead from car exhausts.

(Total 15 marks)
Edexcel

Rivers and estuaries may become polluted by hot water from the cooling towers of power stations.

a) (i) State **two** ways in which thermal pollution may cause the death of aquatic organisms. *(2)*

   (ii) Suggest **two** ways in which this type of thermal pollution may be reduced. *(2)*

b) The sandhopper, *Urothoe*, is a small crustacean, which lives in estuaries. An investigation was carried out into the effects of thermal pollution on its growth.

Two areas in an estuary were sampled regularly over a period of sixteen months (in 1967 and 1968) and the head lengths of the sandhoppers collected in each area were recorded. In one area, hot waste water from a nearby power station entered the estuary, but the other area was unaffected.

The results are shown as a graph in Figure 5.4.

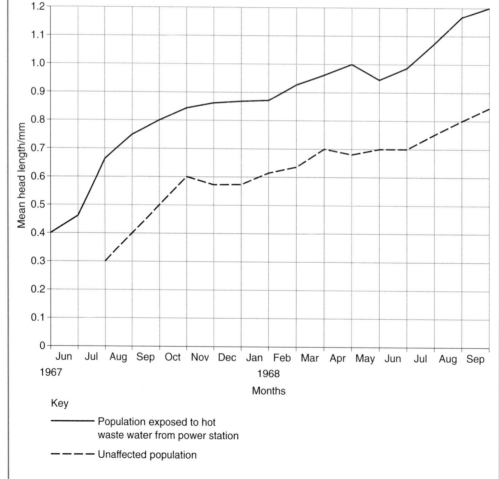

Figure 5.4

EXERCISE 13 continued

    (i)   Calculate the percentage difference between the mean head lengths of the two populations at the beginning of January 1968. Show your working. *(3)*

    (ii) Suggest an explanation for this difference. *(2)*

c)   The graph in Figure 5.5. shows the effect of thermal pollution on the percentage of female sandhoppers carrying eggs from the beginning of March to the end of August 1968. Females carry eggs only during the breeding season.

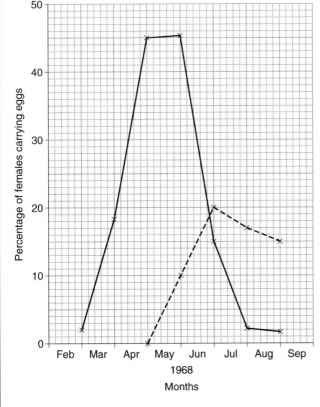

Figure 5.5

    (i)   Comment on the differences between the two curves. *(3)*

    (ii) Suggest why this thermal pollution effect may be a disadvantage to the sandhoppers. *(2)*

(Total 14 marks)

Edexcel

Essays

Question 7

Write an essay on the roles of lipids in living organisms.

Lipids are organic compounds that are insoluble in water but soluble in organic solvents such as ethanol. They are composed mainly of carbon, hydrogen and oxygen, and a number of important biological molecules can be classified as lipids. This essay will describe the general structure of lipids and review the wide variety of roles that they play in living organisms.

The simplest lipids are triglycerides, composed of glycerol joined to three fatty acid molecules. The fatty acids are joined to the glycerol molecule by condensation reactions and the resulting bonds are known as ester bonds. The fatty acids within a particular triglyceride may be identical or different. In general, fatty acids are classified as saturated (containing no carbon/carbon double bonds), e.g. stearic acid, or unsaturated (containing one or more carbon/carbon double bonds), e.g. linoleic acid. It is the different combinations of these fatty acids that give triglycerides their different properties. For example, highly unsaturated triglycerides tend to be liquid at room temperature and are known as oils, whereas saturated triglycerides are solid at room temperature and are known as fats. The main function of triglycerides is to act as energy stores in plants and animals. They are ideally suited for this purpose as they are compact, insoluble and store nearly twice as much energy per gram as carbohydrates. For example, the major energy storage materials of seeds are usually oils.

Another major form of lipids are phospholipids. These have a similar structure to triglycerides, but one of the three fatty acids is replaced by a phosphate group joined to another organic molecule. Phospholipids are the main component of cell membranes. The phosphate 'head' of the molecule is hydrophilic (water soluble) and the fatty acid 'tail' is hydrophobic (insoluble in water). This property of phospholipids means that they form a bilayer in aqueous solutions, which is an important factor in the formation of the membranes.

Lipids have a protective and an insulating role in animals. They are stored around vital organs to cushion them against damage, especially in areas where skeletal protection is lacking, such as the kidneys. Lipid storage in the skin is also important for thermoregulation. This can be shown by looking at the distribution of subcutaneous fat in mammals living in cold environments.

Polar bears, seals and whales all have a thick layer of fatty tissue to keep them warm in arctic conditions. Lipids also act as electrical insulators in the nervous system, where most neurones are surrounded by a myelin sheath. Myelin is composed of a mixture of phospholipids and cholesterol (a steroid) and the myelin sheath insulates the neurone as well as speeding up the conduction of nerve impulses.

There are a number of other important roles of lipids in living organisms. They aid buoyancy in aquatic mammals (as well as providing thermal insulation) and also act as a source of metabolic water (water from respiration) in animals living in arid environments, such as the kangaroo rat. Waxes are a class of lipids that are used for waterproofing in leaves and the exoskeleton of insects. Steroids are another class of lipids that are used as hormones in mammals. For example, the sex hormones testosterone, oestrogen and progesterone are all steroids.

In conclusion, lipids are a very important class of biological molecules. They have a wide range of structures and a variety of functions in both plants and animals.

EXERCISE 14

Write an essay on natural selection and the effects of environmental change.

CHAPTER SIX

MOCK EXAMINATIONS

In this part of the book there are two mock papers, followed by a longer test paper which are written in a similar format to the real unit test papers. When you have completed a paper, ideally under timed conditions, check the mark scheme to see how well you have done. Make sure that you correct any mistakes and that you study the examiner's comments very carefully. You will get a much better grade if you can avoid the common errors made by many candidates in their synoptic unit tests.

Mock papers

PAPER 1
(60 marks: time allowed – $1\frac{1}{4}$ hours)

1   The diagram below shows the structure of a molecule of maltose.

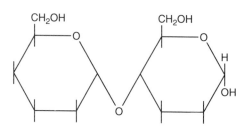

a)   What type of carbohydrate is maltose?   *(1)*

b)   Maltose can be broken down in the gut of mammals to produce glucose. Name the enzyme that catalyses this reaction and its site of action.   *(2)*

c)   A person fasted overnight and then consumed 80 g of glucose. The graph opposite shows the resulting changes in the concentrations of glucose and insulin in the blood of this person.

(i)   Describe the changes in the concentration of glucose in the blood between 30 minutes and 150 minutes.   *(2)*

(ii)   Calculate the percentage increase in the concentration of insulin in the blood between 60 minutes and 90 minutes. Show your working.   *(3)*

(iii) Explain the relationship between the concentrations of glucose and insulin in the blood between 30 minutes and 150 minutes. *(3)*

(iv) Use the information from the graph to explain what is meant by the term *negative feedback.* *(2)*

d) Explain why the concentration of glucagon in the blood rises during exercise, while that of insulin falls. *(2)*

(Total 15 marks)

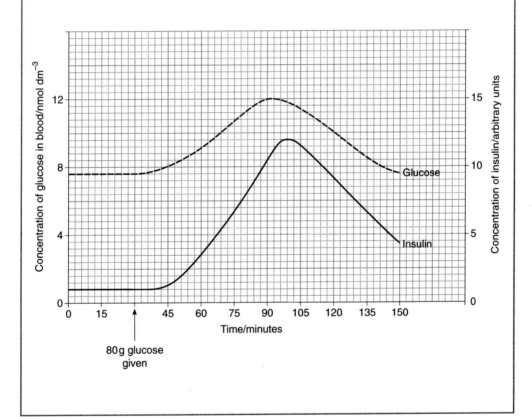

2   Read the passage below and then answer the questions which follow.

C3 plants are plants in which the first product of photosynthesis is a three-carbon compound. In the light-independent stage of photosynthesis, carbon dioxide is accepted by the five-carbon molecule, ribulose bisphosphate. The product of this reaction then immediately splits into two molecules of glycerate 3-phosphate (a three-carbon compound). Examples of C3 plants include most temperate plants, such as wheat and sugar beet.

C4 plants are plants in which the first product of photosynthesis is a four-carbon compound. In the light-independent stage of photosynthesis, carbon dioxide is accepted by the three-carbon molecule, phosphoenolpyruvate (PEP). This forms the four-carbon compound malate. C4 plants represent an adaptation for tropical regions, where the concentration of carbon dioxide during the day is often a limiting factor on the rate of photosynthesis. Malate is mainly produced at night, when there is more carbon dioxide available, and it acts as a store of carbon dioxide. When carbon dioxide is needed during the day, malate is decarboxylated to recycle PEP and release CO_2 for use in the Calvin cycle. The C4 pathway also allows plants to conserve more water and avoids the problem of photorespiration. Examples of C4 plants include many tropical plants, such as maize and sugar cane.

a)   Explain what is meant by the following terms used in the passage:
   (i)   *light-independent stage*   *(3)*
   (ii)   *adaptation*   *(3)*
   (iii) *limiting factor*   *(3)*
   (iv)   *photorespiration.*   *(3)*

b)   The Calvin cycle is a series of biochemical reactions involving triose phosphate and ribulose bisphosphate.
   (i)   Where in a cell does the Calvin cycle take place?   *(1)*
   (ii)   Why is it important to produce ribulose bisphosphate?   *(2)*

   (Total 15 marks)

3 The diagram below shows the single circulation system found in fish.

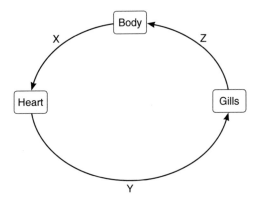

a) At which stage, X, Y or Z, is:
   (i) the pressure at its highest
   (ii) the oxygen concentration at its highest?   *(2)*

b) Compare the single circulation system found in a fish with the double circulation system found in mammals.   *(3)*

A typical electrocardiogram (ECG) for a healthy adult human is known as a PQRST wave, as shown on the graph below.

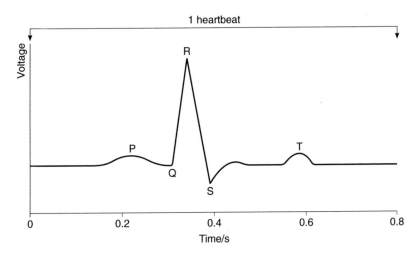

c) Describe what is represented by the following parts of the wave:
   (i) the peak at P
   (ii) the QRS complex
   (iii) the peak at T.   *(3)*

d) Calculate the heart rate in beats per minute of this healthy adult human. Show your working.   *(2)*

PAPER 1 continued

e) The heart rate will increase following stimulation of the sino-atrial node (SAN) by the vagus nerve from the brain. Describe how a nerve impulse is generated and propagated along this nerve. *(4)*

f) The heart rate may also be increased by adrenaline in the blood. State one other effect of adrenaline on the body. *(1)*

(Total 15 marks)

4 Write an essay on the production and elimination of metabolic waste products in plants and animals.

(Total 15 marks)

PAPER 2
(60 marks: time allowed – $1\frac{1}{4}$ hours)

1 a) Explain what is meant by the term *osmosis*. *(2)*

The flowchart below summarises the homeostatic regulation of water potential of human blood .

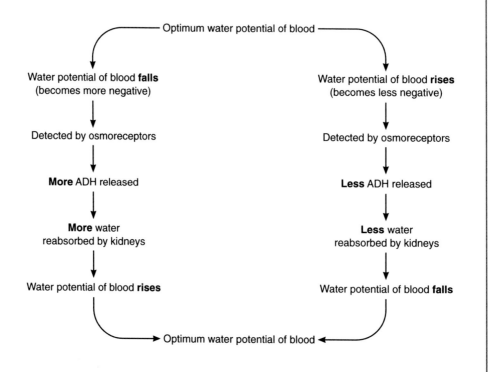

Optimum water potential of blood

Water potential of blood **falls** (becomes more negative)	Water potential of blood **rises** (becomes less negative)
Detected by osmoreceptors	Detected by osmoreceptors
More ADH released	**Less** ADH released
More water reabsorbed by kidneys	**Less** water reabsorbed by kidneys
Water potential of blood **rises**	Water potential of blood **falls**

Optimum water potential of blood

b) (i) Where in the body are the osmoreceptors located? *(1)*

(ii) Name the gland that produces anti-diuretic hormone (ADH). *(1)*

c) Ethanol (alcohol) is a chemical which inhibits the release of ADH. Suggest what would happen to the water potential of the blood of a person who has consumed ethanol. *(3)*

d) The average length of a loop of Henle in an organism is related to the concentration of urine produced according to the formula given below.

$$y = 0.72x + 4$$

Where y = the concentration of urine (arbitrary units); and x = average length of a loop of Henle (mm).

Use this formula to calculate the average length of a loop of Henle in an organism which typically produces urine of concentration 17.5 units. Show your working. *(3)*

e) Animals living in arid (dry) environments tend to have relatively long loops of Henle in their kidneys and high concentrations of ADH in their blood. Suggest an explanation for these observations. *(4)*

(Total 15 marks)

2 The graph below shows how the rates of productivity and respiration change with the leaf area index of tomatoes.

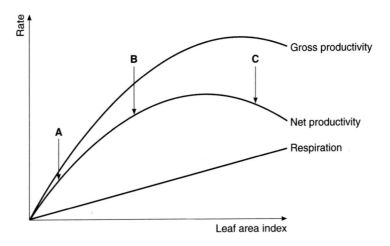

a) (i) State the equation that shows the relationship between gross productivity, net productivity and respiration. *(1)*

(ii) Which of these three quantities will give the best indication of the mass of dry matter produced by the tomatoes? *(1)*

b) Explain the relationship between leaf area index and net productivity between:
   (i) points **A** and **B**   *(2)*
   (ii) points **B** and **C** on the graph.   *(2)*

c) Whitefly is a pest of tomato plants. The table below shows the net productivity of a crop of tomatoes in the presence or absence of whitefly. The leaf area index was identical in both cases.

Condition	Net productivity / arbitrary units
Whitefly present	1420
No whitefly	1680

Calculate the percentage loss of net productivity caused by the whitefly. Show your working.   *(3)*

d) Parasitic wasps can be used as form of biological control of whitefly. An investigation was carried out to find out which of two different species of parasitic wasp would be most effective as a means of biological control. Both species were released into a glasshouse at the same time. The graph below shows the changes in populations of both species of wasp in the period following their release.

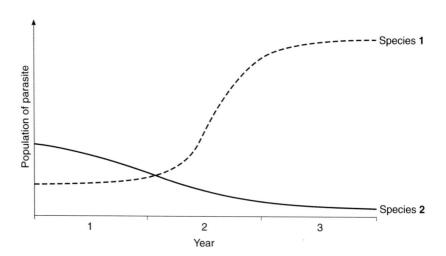

Explain the changes in the population size of Species 1 during this period.   *(4)*

e) Outline **two** advantages of using biological control rather than chemical control to regulate the number of insect pests.   *(2)*

(Total 15 marks)

PAPER 2 continued

3 At high altitudes, atmospheric pressure is lower than at sea level. This means that people living at high altitudes have less oxygen available to them compared with those living at lower altitudes. The table below shows how the percentage saturation of arterial blood with oxygen and the mass of haemoglobin in blood (g per 100 cm³) vary with altitude.

Altitude / m	Percentage saturation of arterial blood with oxygen	Haemoglobin in blood / g per 100 cm³
120	94.8	15.2
1080	91.6	16.6
2250	87.3	17.6
2990	83.9	18.4
4010	80.1	18.8
5250	78.6	18.9

a) (i) Plot the data in a suitable graphical form on graph paper. *(5)*
   (ii) Describe the changes in the mass of haemoglobin in blood as the altitude increases. *(3)*
   (iii) Explain how people living at high altitude have adapted to the low oxygen availability. *(3)*

b) The oxygen dissociation curve for fetal haemoglobin lies to the left of the curve for adult haemoglobin. Suggest an explanation for this difference. *(2)*

c) State **two** ways in which carbon dioxide is transported in the blood. *(2)*

(Total 15 marks)

4 Write an essay on the factors affecting the growth and size of populations.

(Total 15 marks)

The 'real thing'
(120 marks: time allowed – $2\frac{1}{2}$ hours)

1 In an investigation into learned behaviour, a person was asked to draw a line around a metal five-pointed star as quickly as possible. This was done while only looking at the star in a mirror. The number of times contact with the edge of the star was lost was recorded, together with the total time taken to complete the trace around the star. This was repeated ten times, using the same hand each time, with a 10 second rest between attempts.

Attempt	Time taken to complete trace / seconds	Number of errors
1	42	8
2	33	5
3	35	5
4	29	7
5	29	2
6	14	1
7	10	2
8	13	0
9	10	1
10	11	0

a) What conclusions can be drawn from these data? *(2)*

b) A student was asked to design and carry out an experiment to determine if the ability to draw the star is transferred to the hand not used to learn the task. The following account was produced.

The class was divided into two groups, each consisting of five students. All the individuals in one group drew the star once with their right hand, followed by practice with their left hand, before drawing the star again with their right hand. The other group of five students all drew the star twice in succession with their right hand. The average time taken to complete the second star drawn with the right hand was calculated for the two groups, and a statistical test was carried out to see if the difference between the times was significant.

Give **three** criticisms of this experimental design. *(3)*

(Total 5 marks)

AQA (B)

The 'real thing' continued

2 An indicator which turns from red to yellow in the presence of carbon dioxide was used in a class experiment designed to compare the rates of respiration of a range of different living organisms. The time it took to change colour with the same mass of each organism was recorded.

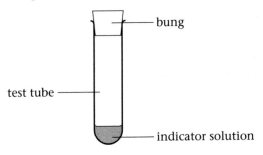

a) (i) What modification would you make to the apparatus to obtain results with blowfly larvae (maggots)? *(1)*

 (ii) Before this apparatus was used to measure the rate of respiration of carrot root, the root was cut into small pieces. How might this have made the results more comparable with those obtained using maggots? *(2)*

b) Before the students introduced organisms into any tube, they were told to take a separate tube containing indicator and blow into it using a straw. This tube was bunged and kept on one side for the duration of the experiment. One student incorrectly described this tube as a 'control'. What was its actual purpose? *(1)*

c) The pooled class results are show in the table below.

Tube	Organisms	Mean time taken for indicator to turn yellow / minutes	Relative rate of respiration
A	Blowfly larvae	15.0	
B	Carrot root	50.0	
C	Mushroom	35.0	
D	Yeast	10.0	
E	Cabbage leaves	no change	

Using the information in the table, calculate values for the relative rates of respiration of organisms A to D, taking the rate of the organism with the slowest rate to be 1.0. *(2)*

d) Students in the class differed in their interpretation of the results obtained with cabbage leaves. Some thought they were caused because the leaves were photosynthesising (hypothesis 1) but others thought that the rate of respiration was too slow to be detected by this experiment (hypothesis 2).

Suggest a way in which you might modify the experiment to test each hypothesis. In **each** case use a different modification, and explain you reasoning in full. *(4)*

(Total 10 marks)

AQA (B)

3   Read the passage and answer the questions which follow.

Red and Grey Squirrels

The red squirrel, *Sciurus vulgaris*, is not as common in Britain as it was a century ago. The grey squirrel, *Sciurus carolinensis*, is now extremely common.

One suggestion for the relative success of the grey squirrel in Britain is that it is able to out-compete the red squirrel. The popular view is that the larger grey squirrels attack the red squirrels. In fact, the two species take very little notice of each other.

Grey squirrels spread slowly and wherever they have been established for more than 15 years, red squirrels were usually missing. Detailed studies have shown that in some areas the two species have coexisted for 16 years or more, but in other areas the reds had disappeared before the greys arrived.

Comparisons of the two species suggest ways in which their ecologies differ so that red squirrels probably do better in coniferous woodland and grey squirrels better in deciduous woods. For instance, reds spend much more time in the canopy and less time on the ground than greys, this matches the fact that they are lighter, more nimble and put on less fat for the winter. Most conifers take two years to ripen their cones, so there are always cones available in the canopy for a squirrel which is light and nimble enough to reach them. The grey squirrel is something of a specialist, feeding mainly on large seeds, such as acorns and beechnuts, that are abundant on the ground in autumn in broad-leaved woodlands. Both squirrels can produce two litters a year; a female grey squirrel, which has the ability to exploit the rich autumn seed crop, will be better placed to produce a strong litter of young early in the year.

Grey squirrels have a further advantage which has probably been decisive. They can digest acorns more efficiently than red squirrels who do eat acorns but they cannot digest them properly. Reds lose weight when given a diet consisting only of acorns. Conservationists have criticised the extensive use of alien conifers by commercial forestry in Britain and have asked for more native conifers and broad-leaved trees to be planted. This is just what grey squirrels prefer. Serious consideration is now being given to felling oaks in areas which are red squirrel strongholds.

Adapted from: *D.W. Yalden, Squirrel Dynamics.*
Biological Sciences Review, Vol. 6 No. 2

a)   State **three** pieces of evidence that support the idea that grey squirrels have not **actively** displaced the native red ones.   *(3)*

b)   Explain whether there is any evidence for the idea that the two species of squirrel occupy the same niche.   *(3)*

The 'real thing' continued

c) Explain why

    (i)   red squirrels are better adapted than grey squirrels to live in coniferous woodland. *(2)*

    (ii) grey squirrels are better adapted than red squirrels to live in deciduous woodland. *(2)*

d) Describe the measures that may be taken to encourage the red squirrel to gain an advantage over the grey squirrel. *(3)*

(Total 13 marks)

OCR

4   The figure below illustrates energy flow diagrams for **A** a deciduous forest and **B** a marine community. The units on the flow charts are kJ m^{-2} day^{-1}. The major plants in the marine community are phytoplankton.

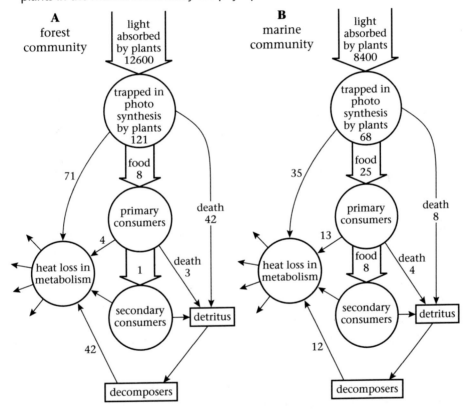

a) Calculate the efficiency with which solar energy is trapped by the forest plants. *(1)*

b) With reference to the diagrams, describe and explain the major differences between the energy flow in these two communities. *(8)*

c) Suggest **two** ways in which the deciduous forest may be managed for timber production. *(2)*

(Total 11 marks)

OCR

The 'real thing' continued

5 Read the following passage.

For a long time it has been known that there are three distinct types of cone cell in the human retina. Each of these possesses a different pigment which absorbs light of a different wavelength. Thus there are cones for red, green and blue. Additionally, genetic studies have enabled us to trace the inheritance of abnormalities in colour vision within families.

Genetic studies have been based on the observation that deficiencies in red and green discrimination are more common in males than in females. Analysis of the pattern of inheritance indicated that genes on the X chromosomes are responsible for the difference. Males will have deficient red discrimination, for example, if the X chromosome, inherited from the mother, carries the relevant allele. Females will only be affected if they receive the allele from both parents. Other genetic studies showed that the difference in blue sensitivity involves a gene on a non-sex chromosome.

Recent work has involved isolating the genes that code for the colour pigments and comparing their structure in people with normal vision with their structure in individuals with deficient colour vision. Frequently genes are isolated by determining the amino acid sequences of the proteins for which they code and using the information as clues to the structure of the genes. Because virtually nothing was known about colour pigment proteins, a less direct approach was used.

Investigators had successfully isolated the rhodopsin from the retina of cattle (bovine rhodopsin). It was planned to isolate the bovine rhodopsin gene and then to use this as a probe to identify the human rhodopsin gene and the cone-pigment genes. The plan relied on a technique known as DNA hybridisation which makes use of the fact that a single strand of DNA will form a stable double helix with a second strand. If the initial strand, known as a probe, is radioactively labelled, then the DNA to which it binds can be identified.

In the second stage of the investigation, a strand of the newly identified bovine rhodopsin gene was used as a probe to search for human rhodopsin and colour pigment genes. The probe bound strongly to only one segment of human DNA but it also bound, but not so strongly, to three other DNA segments. Interestingly, two of these three weakly hybridising genes were located on the X chromosome, while the third gene came from chromosome seven.

Adapted from: *The Genes for Colour Vision*, J. NATHANS, Scientific American, 1989.

The 'real thing' continued

Using information in the passage and your own knowledge, answer the following questions.

a) Explain briefly how the possession of three distinct types of cone cell enables a person to see:

    (i)   red light; *(1)*

    (ii) magenta (purple) light; *(1)*

    (iii) white light. *(1)*

b) (i) What is the evidence from the passage that the allele producing deficient red discrimination is recessive? *(1)*

    (ii) Use a genetic diagram to explain how a female may inherit deficient red discrimination even though her mother had normal colour vision. *(3)*

c) On which chromosome is the gene controlling blue sensitivity located? Explain your answer. *(2)*

d) In terms of DNA structure, explain how single stranded DNA can form a 'stable double helix with a second strand'. *(2)*

e) Explain why the bovine rhodopsin gene binds:

    (i) 'strongly to only one segment of human DNA'; *(2)*

    (ii) 'not so strongly to three other DNA segments'. *(2)*

(Total 15 marks)

AQA (A)

6 The liver is unusual in having a double blood supply, from the hepatic portal vein and the hepatic artery. The liver receives over 1 dm³ of blood every minute when the body is at rest.

a) Suggest why the liver has a double blood supply. *(2)*

b) Outline the role of the liver in the control of blood glucose.

(In this question, 1 mark is available for the quality of written communication.) *(9)*

c) (i) Explain why it is important that the liver regulates blood cholesterol. *(4)*

    (ii) State **two** possible consequences of having a high blood cholesterol concentration. *(2)*

d) Suggest why a person suffering from liver damage is often prescribed smaller doses of any medication required for other complaints. *(2)*

Jaundice is frequently an indication of liver damage. In this condition, the skin takes on abnormal yellow colouration because bile pigments are circulating in the blood.

e) Explain why the bile pigments are circulating in the blood of a person suffering from jaundice. *(2)*

(Total 21 marks)

OCR

7 **a)** Microorganisms present in a rabbit's gut are able to digest carbohydrates in the plant material that they eat. Figure 1 shows the biochemical pathways by which cellulose and starch are digested in the gut of a rabbit.

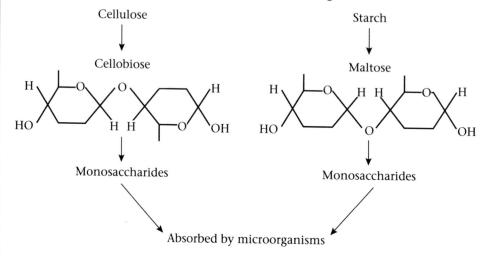

FIGURE 1

(i) Describe how a molecule of cellulose differs from a molecule of starch. *(1)*

(ii) Draw a diagram to show the molecules produced by digestion of cellulose. *(2)*

(iii) Cellobiose and maltose are both disaccharides. Explain why amylase enzymes produced by the rabbit are unable to digest cellobiose. *(3)*

b) One way in which rabbits cause considerable damage to agricultural land is by competing for plant material that would normally be eaten by domestic animals. Table 1 shows some features of the energy budgets of rabbits and cattle living under the same environmental conditions. All figures are kilojoules per day per kilogram of body mass.

	Rabbits	Cattle
Energy consumed in food	1272	424
Energy lost as heat	567	311
Energy gained in body mass	68	17

TABLE 1

(i) What is the purpose of giving these figures per kilogram of body mass? *(1)*

(ii) Explain the difference in the figures for the amount of energy lost as heat. *(2)*

(iii) Use the information in Figure 1 to explain why all the energy consumed in food cannot be converted to body mass or is lost as heat. *(2)*

The 'real thing' continued

c) Rabbits were introduced to Australia in the middle of the last century. Their population grew rapidly and they are now major agricultural pests.

Table 2 compares some features concerned with heat loss in cattle and rabbits at a temperature of 30 °C.

	Cattle	Rabbits
Percentage of body heat which is lost by evaporation	81.0	17.0
Core temperature of body	38.2	39.3

TABLE 2

Use the information given in parts (b) and (c) of this question to explain:

(i) how evaporation helps cattle to maintain a constant body temperature; *(2)*

(ii) the main way in which a rabbit would lose heat at an environmental temperature of 30 °C; *(2)*

(iii) why rabbits are major agricultural pests in Australia; *(2)*

(iv) why rabbits are better able to survive than cattle in the hot, dry conditions found in many parts of Australia. *(3)*

(Total 20 marks)

AQA (A)

8 Write an essay on **one** of the topics below.

EITHER

a) The process of diffusion and its importance in living organisms.

OR

b) Mutation and its consequence.

(Total 25 marks)

AQA (B)

ANSWERS

MAKING CONNECTIONS: *Answers to Exercises*

Exercise 1 a): *Spider diagram for enzymes*

(Note that ; in a markscheme indicates a marking point and / indicates alternative correct responses.)

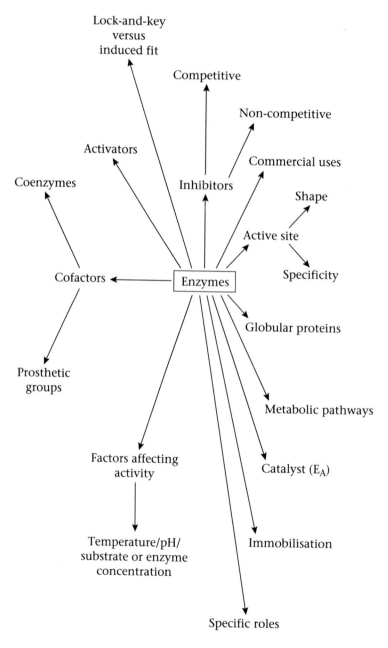

Exercise 1 b): Spider diagram for gas exchange in plants and animals

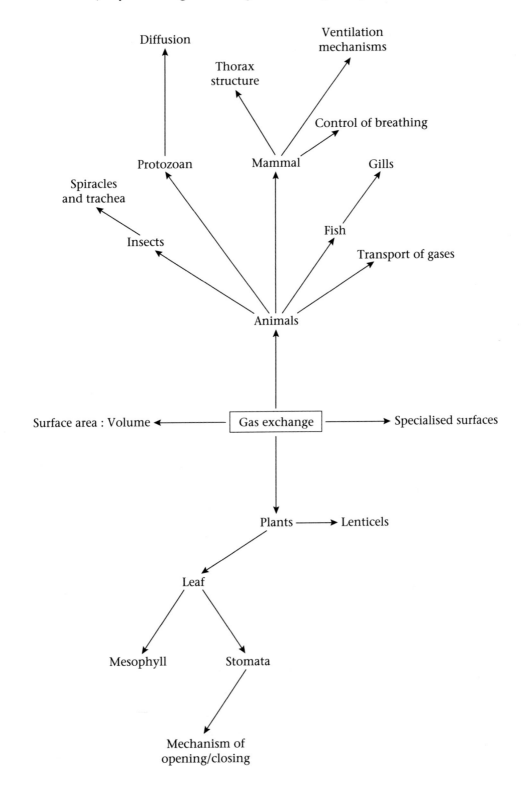

Exercise 2: The effect of altitude on the number of red blood cells in men and women.

Altitude	Gender	Mean number of red blood cells ($\times 10^{12}$) per dm^3
Low	Men	5.29
	Women	4.69
High	Men	5.65
	Women	4.97

Exercise 3

The mistakes made in the graph are:

- no title
- no label (or units) on the *y*-axis
- no units on the *x*-axis
- inappropriate scale on the *y*-axis, so the graph is 'squashed in' at the bottom.

The graph would be much better drawn as shown below.

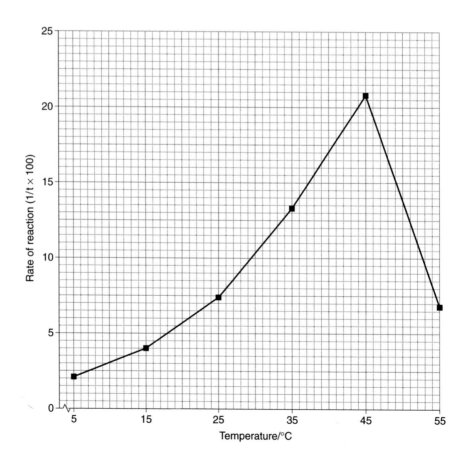

Exercise 4

a) (i) 3400 (ii) 4900 (iii) 4900 *(2)*

b) (i) both have the same total intake/output; both lose the same volume of water from the lungs/in faeces; more water lost in urine on cold day (or converse for warm day); less water lost as sweat on cold day (or converse for warm day); (Any 3) *(3)*
    (ii) more water lost as sweat on race day; due to increased sweating during race to enable heat loss; more water lost from lungs on race day; due to greater rate of ventilation during race; (slightly) more water lost as urine on race day; probably related to greater intake in drinks; (Any 3) *(3)*

c) (i) respiration produces carbon dioxide and water; as oxygen is combined with hydrogen during oxidative phosphorylation/electron transport chain; *(2)*
    (ii) muscles are using more energy/ATP; so there is more respiration/increased metabolism; more glucose is used; (Any 2) *(2)*

d) glycogen is converted to glucose; so the concentration of blood glucose increases; SA node stimulated; so rate of heartbeat increases/greater blood supply to skeletal muscles; (Any 2 linked points) *(2)*

(Total 14 marks)

Exercise 5

a) (i) a weak electrostatic attraction; between a hydrogen atom and an oxygen atom; *(2)*
    (ii) the monomer from which nucleic acids are formed; consisting of a pentose sugar/deoxyribose, a nitrogenous base and a phosphate group; *(2)*
    (iii) a reaction in which two molecules are joined together; and a molecule of water is removed; *(2)*

b) (i) three divisions = 8 × original mass; = 8 × 10 = 80 μg; *(2)*
    (ii) percentage parental DNA = (10/80) × 100; = 12.5%; *(2)*

c) DNA replication involves both polynucleotide strands, transcription involves only one (the 'sense' strand); DNA replication involves only DNA nucleotides, transcription involves both DNA and RNA nucleotides; the end product of DNA replication is DNA, the end product of transcription is messenger RNA (mRNA); the whole DNA strand is copied during DNA replication, only part of the DNA (a gene) is used in transcription; (Any 2 comparative points) *(2)*

(Total 12 marks)

Exercise 6

ESSAY PLAN TEMPLATE

Title: ATP and its roles in living organisms

Introduction: The nature of ATP and its importance as energy currency in cells.

Paragraph 1: Production of ATP in cytoplasm and mitochondria; aerobic respiration.

Paragraph 2: Anaerobic respiration; role of chloroplasts in ATP production.

Paragraph 3: Uses – active transport; maintaining resting potential; reabsorption in nephron; absorption in the gut.

Paragraph 4: Uses – Calvin cycle; muscle contraction; biosynthesis; contractile vacuoles.

Conclusion: ATP is involved in a large number of anabolic reactions and physiological processes.

Exercise 7

Scientific content (S): the essay covers most of the material that might be expected. Some of the areas are explained well, such as specific heat capacity, fertilisation and the role of water in photosynthesis. However, there is too little development of several of the topics raised, e.g. vague references to water as a lubricant, no explanation of turgidity and an inadequate explanation of sweating. Furthermore, there are one or two confused references, such as dissolving antibodies and bacteria. *(Mark: 10/12)*

Balance (B): all of the main areas are addressed and there are references to both plants and animals. There are few errors and no irrelevant points. *(Mark: 2/2)*

Coherence (C): spelling, punctuation and grammar are reasonably sound (e.g. note the mis-spelling of 'capillaries' and 'cerebrospinal fluid') and technical terms are generally used appropriately. There is an attempt at an introduction and a conclusion. However, the material is not logically presented, with a fairly disjointed structure and some repetition. Some of the paragraphs could have been amalgamated and put into a more logical order. *(Mark:1/2)*

(Total mark: 10 + 2 + 1 = 13/16)

Exercise 8

a) (i)

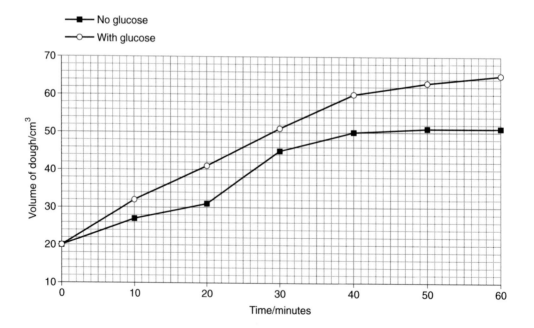

(5)

(ii) mean rate of change with glucose = $[(36.5 - 20)/15] \times 60$; = $66 \, cm^3 \, h^{-1}$; so difference = $66 - 36 = 30 \, cm^3 \, h^{-1}$; *(3)*

(iii) the dough rises faster with glucose; volume continues increasing with glucose, but levels off without glucose; *(2)*

(iv) glucose is a respiratory substrate; so more carbon dioxide is produced; *(2)*

b)   the volume will increase more rapidly; because amylase breaks down starch to glucose; this provides more substrates for respiration; so more CO_2 produced; initial rate will not be as fast as in dough with glucose added; (Any 3)   *(3)*

(Total 15 marks)

Exercise 9

a)   (i)   to make the drops visible when added to beaker A;   *(1)*

(ii) to show that any differences were due to the presence of the tissue;   *(1)*

(iii) to remove any sap from the cells that might contaminate the solutions;   *(1)*

b)   if the water potential of the tissue was lower than the surrounding solution, then water is gained from the solution by osmosis; this makes the external solution more dense than the original solution; so the drops are more dense and they sink in the solution; if the water potential of the tissue is the same as the solution, the drops will remain suspended; (accept converse argument for drops that rise)   *(4)*

c)   the water potential of the potato tissue is equivalent to $0.15 \, mol \, dm^{-3}$ sucrose solution; the water potential of the beetroot tissue is equivalent to $0.25 \, mol \, dm^{-3}$ sucrose solution; so the water potential of beetroot tissue is lower than that of potato tissue; (Any 2)   *(2)*

d)   sucrose is soluble and starch is not; so sucrose affects the solute potential of the cell and starch does not;   *(2)*

e)   replication; same number of slices in each tube; same size/surface area of slices; from same potato/beetroot; tubes kept at same temperature; same volume of solution in each tube; mix before removing drops; use clean syringe each time; (Any 3)   *(3)*

(Total 14 marks)

Exercise 10

a)   (i)   Arthropoda;   *(1)*

(ii) exoskeleton; jointed legs;   *(2)*

b)   (i)   as caffeine concentration increases, heart rate increases (or converse); not a linear relationship/greater effects at higher concentrations; no difference in rate below 0.0001 parts per million (inclusive)/any correct manipulation of figures;   *(3)*

(ii) $[(5.6 - 3.7)]/100; = 51.35\%$;   *(2)*

c)   (i)   to find heart rate without caffeine/compare treated and untreated *Daphnia*;   *(1)*

(ii) to enable calculation of a mean heart rate; increases the reliability of the experiment;   *(2)*

(iii) *Daphnia* are ectotherms; changes in temperature will affect metabolism; changes in temperature will affect heart rate; important to keep such extraneous variables constant; (Any 3)   *(3)*

(Total 14 marks)

Exercise 11

a) binds with receptor; opens sodium channels; sodium ions enter the neurone; membrane is depolarised; (Any 3) *(3)*

b) dorsal root ganglia contain cell bodies of sensory neurones; only these neurones produce the capsaicin receptor; *(2)*

c) only some DNA is transcibed to mRNA; so fewer different types of mRNA in cells; easier to find required sequence; (Any 2) *(2)*

d) monomers/sub-units of a DNA molecule; containing deoxyribose, base and phosphate group; *(2)*

e) reverse transcriptase; *(1)*

f) find nucleotide sequence of cDNA; three nucleotides/bases code for one amino acid; use genetic code to identify amino acids from bases; *(3)*

g) exposure to capsaicin reduces sensitivity; of neurones which respond to inflammation; so fewer action potentials along these neurones; (Any 2) *(2)*

(Total 15 marks)

Exercise 12

a) (i) 2.08 – 2.1 ppm; *(1)*
   (ii) resistant mosquitos are likely to survive; and will breed, leading to increased resistance in the population; *(2)*

b) set up larvae without insecticide; to find natural mortality rate/effect of insecticide; *(2)*

c) may be carried into rivers and lakes; may harm other living organisms; may bioaccumulate because it is non-biodegradeable; *(3)*

d) acetylcholine is a neurotransmitter; insecticide prevents removal of acetylcholine at synapses; interferes with nerve impulse transmission; causing abnormal muscle function; (Any 3) *(3)*

e) (i) insects unlikely to become resistant; natural predators are more specific; no adverse environmental effects with natural predators, e.g. bioaccumulation; no need to reapply treatment as natural predators reproduce; (Any 2) *(2)*
   (ii) will not remove all the pest; slow acting; predator may become a pest; (Any 1) *(1)*

(Total 14 marks)

Exercise 13

a) (i) lack of oxygen due to lower solubility/increased decomposition; denaturation of enzymes; loss of food species; (Any 2) *(2)*
   (ii) allow water to cool before returning; transfer waste heat to air; use heat for homes/glasshouses/fish farms; use alternative sources of energy; (Any 2) *(2)*

b) (i) $[(0.87 - 0.58); /0.87] \times 100; = 33.3\%;$
       OR $[(0.87 - 0.58); /0.58] \times 100; = 50\%;$ *(2)*
   (ii) increase in temperature increases rate of growth; more food available for growth; increase in temperature leads to higher metabolic rate; (Any 2) *(2)*

c) (i) population exposed to hot water starts breeding season earlier in year; peak lasts longer; maximum percentage of females carrying eggs is higher; appropriate use of figures; (Any 3) *(2)*
   (ii) hatching time of larvae may not coincide with maximum food availability; may be greater numbers of predators; increased numbers leads to greater intraspecific competition; (Any 2) *(2)*

(Total 14 marks)

Exercise 14

This essay should be balanced and coherent (see page 26), and include descriptions and discussion within all or most of the following areas.

Natural selection

- variety amongst individuals of same species
- an individual has many different characteristics that affect its survival
- some characteristics give individuals an advantage
- reference to genes/alleles
- more likely to survive and breed
- offspring inherit the characteristics
- characteristics become more common in the population
- certain characteristics are selected for/against.

Environment and selection pressures

- environmental factors affect survival/exert selection pressure
- examples of environmental factors, such as change in climate, pollution, predation, food availability
- changes in environment will favour certain characteristics
- reference to differential mortality/natality
- stabilising selection
- directional selection
- disruptive selection
- leads to isolation and speciation.

Specific examples

- industrial melanism in peppered moth
- heavy metal tolerance in plants
- insecticide/antibiotic resistance.

MOCK EXAMINATIONS:
Mark Schemes and Examiner's Comments

PAPER 1

1 a) disaccharide; *(1)*

b) Name: maltase; Site of action: ileum/small intestine; *(2)*

Examiner's Tip:
There are a number of important digestive enzymes and you should know where each one is produced, its substrates and products, and the conditions under which it works best.

c) (i) glucose concentration rises steadily from 30 minutes to 90 minutes; it then falls back to its original level at 150 minutes; *(2)*

Examiner's Tip:
Where figures appear in a question (as they do in this graph), you should try to use them. For example, in this case you could say that the glucose concentration rises steadily from 7.6 nmol dm^{-3} at 30 minutes to 12 nmol dm^{-3} at 90 minutes.

(ii) percentage increase = [(10.5 − 3.5); /3.5] × 100; = 200%; *(3)*

Examiner's Tip:
Remember that PERCENTAGE CHANGE = (CHANGE/ORGINAL) × 100. Note that you can have a percentage increase of more than 100%. Some students go back and change their response if they get more than 100%. Make sure that you avoid this error!

(iii) increase in glucose concentration stimulates the release of insulin; as glucose concentration continues to rise (due to absorption by the blood), more insulin is released; insulin eventually causes concentration of glucose to fall; due to stimulating glucose uptake by liver and muscle cells; this causes a decrease in insulin secretion (due to less glucose in the blood); (Any 3) *(3)*

Examiner's Tip:
If 3 marks are available for a question you should ensure that you make three separate points.

(iv) negative feedback mechanisms maintain physiological factors at their optimum; if blood glucose rises, insulin is secreted which causes the glucose level to fall back to the optimum; *(2)*

Examiner's Tip:
*Note that in cases of **positive feedback**, a deviation from the optimum causes a greater deviation to occur.*

d) exercise reduces blood glucose concentration; due to more respiration/greater metabolic rate; stimulates the secretion of glucagon/inhibits the secretion of insulin; *(2)*

(Total 15 marks)

2 a) (i) stage of photosynthesis that does not require light; takes place in stroma; carbon dioxide is accepted by a three- or five-carbon compound; then reduced to carbohydrate; (Any 3) *(3)*

Examiner's Tip:
*Try not to confuse this process with the **light-dependent stage,** which requires light and involves the production of ATP, reduced NADP and oxygen.*

(ii) any feature of the structure/physiology of an organism; that makes it well suited to its environment; so it is more likely to survive and reproduce (than organisms that are less well adapted); example: palisade cells are adapted for efficient photosynthesis/ xerophytes are adapted to conserve water in arid environments/polar bears are adapted to live in cold climates; (Any 3) *(3)*

(iii) any environmental factor; that limits the rate of a biological process; usually the factor in least supply; example: light intensity/temperature (**not** CO_2) limiting rate of photosynthesis; (Any 3) *(3)*

(iv) at high concentrations of oxygen; oxygen competes with CO_2 for the active site on ribulose bisphosphate carboxylase; so oxygen combines with ribulose bisphosphate; less CO_2 is fixed/less carbohydrate is formed; (Any 3) *(3)*

b) (i) in the stroma (of chloroplasts); *(1)*

Examiner's Tip:
When asked a 'where' question, try to be as specific as possible. The answers 'leaf', 'palisade cell' and 'chloroplast' would not be incorrect for this question, but would not be specific enough to earn the mark.

(ii) ribulose bisphosphate in the CO_2 acceptor in C3 plants; without it, no CO_2 would be fixed and photosynthesis would stop; plants would die/disruption of food chain; (Any 2) *(2)*

Examiner's Tip:
Sometimes a question like this will ask you 'what is the significance of ...?'.
This means 'why is ... important for the organism?'.

(Total 15 marks)

3 a) (i) Y; *(1)*

(ii) Z; *(1)*

Examiner's Tip:
Remember that the pressure will be highest just after leaving the heart (Y) and the oxygen concentration will be highest just after leaving the gills (Z).

b) single circulation: blood passes through heart once as it travels around body; not efficient as much pressure lost passing through gas exchange surface (gills); double circulation: blood passes through heart twice as it travels around body; more efficient as pumped again after passing through gas-exchange surface (lungs); (Any 3) *(3)*

c) (i) P = atria contracting/atrial systole; *(1)*
(ii) QRS = ventricles contracting/ventricular systole; *(1)*
(iii) T = atria and ventricles relaxing/atrial and ventricular diastole; *(1)*

Examiner's Tip:
Students sometimes mix up the sequence of events in the cardiac cycle. Remember that the Atria contract before the Ventricles (A is before V in the alphabet).

d) one heart beat = 0.8 seconds; so heart rate = (60/0.8) = 75 beats per minute; *(2)*

Examiner's Tip:
*In calculations such as these, you **expect** an answer in the typical range for heart rate, e.g. 60–90 beats per minute. If you had made a mistake in the calculation, e.g. 60 × 0.8 = 48 beats per minute, you should notice the error!*

e) nerve cell originally at resting potential; a high concentration of sodium ions outside the cell/inside negative relative to outside/typical resting potential of −70 mV; stimulation of nerve cell leads to depolarisation; sodium ions enter cell/inside becomes positive relative to outside/action potential; propagated by wave of depolarisation along length of nerve cell; possible saltatory conduction along myelinated nerve fibres; (Any 4) *(4)*

f) stimulates conversion of glycogen to glucose/stimulates metabolic rate; *(1)*

(Total 15 marks)

4 This essay should be balanced and coherent (see page 26), and include descriptions and discussion within all or most of the following areas.

General points

- requirement for removal of toxic waste products of metabolism
- removal may involve special metabolic pathways and/or specialised organs.

Carbon dioxide

- waste product of aerobic respiration
- diffuses out of stomata/lenticels in plants
- diffuses into tissue fluid/blood plasma of vertebrates
- details of transport as hydrogen carbonate/carbamino compounds
- reference to Bohr shift
- diffusion into alveoli of lungs from pulmonary capillaries
- expiration mechanism
- reference to control of ventilation by medulla in brain
- elimination of carbon dioxide in other animals, e.g. insects.

Nitrogenous waste

- produced by deamination of excess amino acids
- formation of urea from ammonia (via ornithine cycle) in liver
- urea transported to kidneys via plasma
- ultrafiltration and removal of urea (as urine) by kidneys
- nitrogenous waste removal in other animals, e.g. uric acid in insects.

Other

- reference to oxygen production in photosynthesis in plants
- metabolic waste removal via leaf abscission, e.g. tannins.

(Total 15 marks)

PAPER 2

1 a) the diffusion of water molecules; across a partially permeable membrane; from an area of high water potential to an area of low water potential/down a water potential gradient; *(3)*

Examiner's Tip:
It is important to know the precise definitions of biological terms and it may be useful to buy a dictionary for AS/A-level Biology.

b) (i)   hypothalamus; *(1)*
(ii) pituitary gland; *(1)*

c) inhibition of ADH release means less water reabsorbed from the nephrons back into the blood; more urine produced/the water content of the blood falls; blood has a lower (more negative) water potential; *(3)*

d) $17.5 = 0.72x + 4$;
$0.72x = (17.5 - 4) = 13.5$;
$x = 13.5/0.72 = 18.75$ mm; *(3)*

Examiner's Tip:
Always show your working in calculations – you might earn some marks even if your final answer is wrong! Don't forget to put units in the answer.

e) loop of Henle acts as a countercurrent multiplier/builds up concentration of sodium in the medulla; this results in the reabsorption of water by osmosis/the production of concentrated urine in the collecting duct; high concentrations of ADH make collecting ducts more permeable to water; so a lot of water is reabsorbed back into the blood from the kidney nephrons; adaptive mechanisms help the animals to conserve water in dry environments; (Any 4) *(4)*

(Total 15 marks)

2 a) (i)   net productivity = gross productivity – respiration; *(1)*

Examiner's Tip:
*Remember that **net** productivity is a measure of the rate at which energy is used to form new tissues. Therefore it can be calculated from **gross** productivity (total amount of material accumulated) less respiration (material used to form energy).*

(ii) net productivity; *(1)*

b) (i)   as leaf area index increases, there is a greater surface area for photosynthesis; more organic products of photosynthesis; *(2)*

Examiner's Tip:
*Note that this is an **explain** question and you would not get any marks for simply **describing** the relationship (see page 19).*

(ii) at high leaf area index the leaves begin to shade each other/lower leaves are shaded by upper leaves; therefore there is less photosynthesis **per unit leaf area**; but leaves will still respire; decreased ratio of photosynthesis to respiration; (Any 2) *(2)*

c) percentage loss = [(1680 – 1420); /1680] × 100; = 15.5%; *(3)*

d) interspecific competition; is limiting factor on population of Species 1 for first two years; Species 1 better adapted/more successful than Species 2 so numbers rise rapidly (in year 2); population of Species 1 levels off in year 3 due to intraspecific competition/limited numbers of whitefly hosts; *(4)*

Synoptic Skills in Advanced Biology

Examiner's Tip:
*When asked about factors affecting the distribution or numbers in a population, always try to consider both **abiotic** factors (non-living, e.g. pH or oxygen availability) and **biotic** factors (living, e.g. predator-prey relationships).*

e) biological control is more specific; less chance of resistance; no adverse environmental effects, e.g. bioaccumulation; no need to reapply, as natural predators reproduce; (Any 2) *(2)*

(Total 15 marks)

3 a) (i)

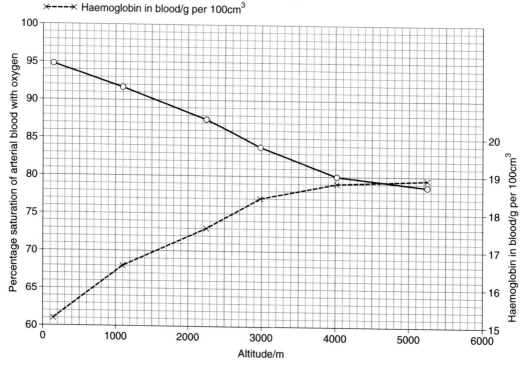

(5)

Examiner's Tip:
Remember the SPACK mnemonic given on page 15.

(ii) it increases as altitude increases; increase at higher altitudes/over 3000 m is less than at lower altitudes/less than 3000 m; manipulation of figures, e.g. increase of 24.3% between 120 m and 5250 m; *(3)*

Examiner's Tip:
Remember that you will not get any marks for simply restating the figures from the table – you need to manipulate the data in some way, i.e. do a simple calculation.

(iii) haemoglobin transports oxygen in blood; so more haemoglobin means more O_2 carried; to compensate for lower percentage saturation of arterial blood; other valid physiological changes, e.g. greater lung capacity/increased cardiac output/more red blood cells; *(3)*

b) fetal haemoglobin has a different structure from adult haemoglobin; it has a higher affinity for oxygen; so that it can absorb oxygen from the maternal circulation/across the placenta; (Any 2) *(2)*

Examiner's Tip:
*If the oxygen dissociation curve moves to the **left**, haemoglobin will have a **higher** affinity for oxygen. Likewise, if it moves to the **right**, haemoglobin has a **lower** affinity for oxygen. Try to remember that a shift to the **Right** makes it more likely that haemoglobin will **Release** oxygen.*

c) as hydrogen carbonate ions (in plasma); as carbamino compounds/bound to haemoglobin; *(2)*

(Total 15 marks)

4 This essay should be balanced and coherent (see page 26), and include descriptions and discussion within all or most of the following areas.

General points

- population defined
- description and explanation of typical population growth curve
- carrying capacity
- population change = (births + immigration) − (deaths + emigration)
- density-dependent and density-independent factors.

Abiotic factors

- examples (light, inorganic ions, oxygen, temperature, etc.)
- competition for abiotic factors.

Biotic factors

- food supply
- spread of disease/parasitism
- predator-prey relationships
- interspecific competition
- intraspecific competition
- competitive exclusion.

Other points

- demographic changes in human populations/population pyramids
- birth control.

(Total 15 marks)

The 'real thing'

1 a) Time taken to complete trace and number of errors decreases; until no further improvement made after 7 attempts; *(2)*

b) does not test hypothesis; sample size too small; age/gender not considered; not random or right-handed/left-handed not considered; period of practice not considered; same time needed between attempts for both groups; *(3)*

(Total 5 marks)

2 a) (i) e.g. metal grid/cage holding maggots above liquid; *(1)*

(ii) similar surface area; for gaseous exchange *(2)*

b) colour standard to judge endpoint *(1)*

c) 3.3, 1, 1.4, 5 *(2)*

d) Hypothesis 1 Cover tube with e.g. metal foil; this will give dark conditions where cabbage leaves cannot photosynthesise

Hypothesis 2 Place tube in water bath at 30°C; increased temperature will increase rate of reaction of respiratory enzymes *(4)*

(Total 10 marks)

3 a) can live together for 16 years; ignore each other; in some areas reds disappeared before greys arrived; *(3)*

b) no evidence; reds, coniferous woods and greys, deciduous woods; grey feeds on large seeds, e.g. acorns; red feeds on small seeds, e.g. conifer seeds; red spends more time in the canopy than grey; coexist together in same areas; *(3)*

c) (i) light; nimble; put on less fat in the winter; *(2)*

(ii) can digest acorns; lots of acorns in autumn increases body reserves; increased breeding success; *(2)*

d) cull/kill, greys; feed reds; plant trees with small seeds; remove oaks; use contraceptives on greys; *(3)*

(Total 13 marks)

4 a) 0.96% *(1)*

b) any two valid differences between the energy flow in A and B supported by figures; ref to differences between leaves and phytoplankton in absorbing light; ref to limiting factors of photosynthesis and effects on energy flow in two communities; higher energy flow through consumer food chain in B; higher energy flow through detritus food chain in A; larger population of decomposers in forest; inedible material produced in forest, e.g. wood; smaller, number/biomass of secondary consumers in A; ref to efficiency of transfer of energy between trophic levels; AVP *(8)*

c) large trees felled, replanted; coppicing, described; *(2)*

(Total 11 marks)

5 **a)** (i)   red cones only stimulated;   *(1)*

   (ii)   red and blue cones stimulated;   *(1)*

   (iii) all cones stimulated;   *(1)*

b) (i)   females need allele from both parents before they are affected;   *(1)*

   (ii)   phenotypes of parents: female normal vision × male colour-blind
            genotypes of parents: X^RX^r X^rY ;
            gametes X^R X^r X^r Y
            offspring genotypes X^RX^r X^RY X^rX^r X^rY ;
            X^rX^r identified as a colour-blind female;   *(3)*

c) chromosome 7; gene is on non-sex chromosome and one of the colour genes is on this chromosome;   *(2)*

d) if there is a complementary base sequence; hydrogen bonds can be formed;   *(2)*

e) general principles – more similar the proteins, the more bases in common; link between bases, hydrogen bonds and stability;

   (i)   bovine rhodopsin and human rhodopsin very similar proteins;

   (ii)   less in common with the proteins in colour pigments;   *(4)*

(Total 15 marks)

6 **a)** hepatic artery supplies liver with oxygen/oxygenated blood; hepatic portal vein brings products of digestion/nutrients; ref to homeostatic role of liver   *(2)*

b) when blood glucose level rises, insulin secreted; causes conversion of glucose to glycogen/glycogenesis; <u>increased</u> respiration/uptake of glucose in liver cells; when blood glucose level falls, glucagon secreted; causes conversion of glycogen to glucose/glucogenesis; production of glucose from non-carbohydrate/ gluconeogenesis; named non-carbohydrate source; adrenaline secretion causes conversion of glycogen to glucose; ref islets of Langerhans; α, β cells/pancreas, in secreting hormones/detecting blood glucose; ref to homeostasis; ref negative feedback;   *(8)*

(1 mark for a clear, well organised answer using specialist terms)

c) (i)   available to cells; maintain constant blood concentration; for membrane synthesis; ref to role of cholesterol in cell membranes; for steroid/hormone synthesis; ref to vitamin D synthesis;   *(4)*

   (ii)   gall stones; atherosclerosis; CHD; stroke;   *(2)*

d) liver processes/detoxifies/eliminates drugs; drug may build up in body/not broken down as quickly; increased side/toxic/harmful effects if full dose given; therapeutic/beneficial effects will last longer;   *(2)*

e) haemoglobin broken down/converted to bile pigments; ref to bilirubin and biliverdin; bile pigments not excreted; because liver cells not able to remove from blood;   *(2)*

(Total 21 marks)

7 **a)** (i)   cellulose is made from β-glucose and starch from α-glucose;   *(1)*

   (ii)   diagram showing two recognisable monosaccharides; correct groups produced as result of hydrolysis;   *(2)*

   (iii) cellobiose and maltose have different shapes; cellobiose will not fit/bind to/form enzyme-substrate complex with; active site of amylase;   *(3)*

b) (i)   to take into account the difference in size of the animals; *(1)*

        (ii) heat is lost from surface; rabbit is smaller and has a greater surface area-volume ratio; *(2)*

       (iii) some of the energy consumed in food is used by microorganisms; for respiration or microbial growth; *(2)*

    c) (i)   evaporation requires (latent) heat; to convert water to vapour; this heat is drawn from the animal's body; *(2)*

        (ii) must lose it directly to cooler environment; by radiation/convection; *(2)*

       (iii) rabbits not kept in check by natural predators/no predators of rabbit in Australia; consume much more food per kilogram than cattle; competition for limited resources; *(2)*

       (iv) rabbit has a higher body temperature so does not have to lose as much heat; can lose more heat to environment by radiation; does not rely as much on sweating as cattle; less dependent on drinking water; *(3)*

                                                              (Total 20 marks)

8   Four skill areas will be marked scientific content (max 16 marks), breadth of knowledge (max 3 marks), relevance (max 3 marks) and quality of written communication (max 3 marks).

                                                                      (Total 25 marks)